Henry Gardiner Adams

The Young Naturalist

A Handy Volume Upon the Collection, Preservation, and Arrangement...

Henry Gardiner Adams

The Young Naturalist
A Handy Volume Upon the Collection, Preservation, and Arrangement...

ISBN/EAN: 9783337025823

Printed in Europe, USA, Canada, Australia, Japan

Cover: Foto ©berggeist007 / pixelio.de

More available books at **www.hansebooks.com**

1. Greasy Fritillary.   2. Glanville Fritillary.   3. Pearl-bordered Fritillary.
4. Weaver's Fritillary.

THE

YOUNG NATURALIST

A

HANDY VOLUME

UPON THE

COLLECTION, PRESERVATION, AND ARRANGEMENT

OF

BUTTERFLIES AND SHELLS.

BY

H. G. ADAMS.

Illustrated.

LONDON:
GROOMBRIDGE AND SONS.
1879.

BEAUTIFUL BUTTERFLIES.

BEAUTIFUL BUTTERFLIES

Described and Illustrated

WITH THE

HISTORY OF A BUTTERFLY

THROUGH ALL ITS CHANGES AND TRANSFORMATIONS;

AND AN EXPLANATION OF THE

Scientific Terms used by Naturalists in reference thereto.

By H. G. ADAMS,

AUTHOR OF 'NESTS AND EGGS OF FAMILIAR BIRDS,' 'FAVORITE SONG BIRDS,'
'BEAUTIFUL SHELLS,' 'HUMMING BIRDS,' ETC. ETC.

ILLUSTRATED WITH COLOURED PLATES AND
NUMEROUS WOOD ENGRAVINGS.

LONDON:
GROOMBRIDGE AND SONS.

PRINTED BY J. E. ADLARD,
BARTHOLOMEW CLOSE.

CONTENTS.

DESCRIPTION OF SPECIES (*continued*)

BEAUTIFUL BUTTERFLIES.

INTRODUCTION.

"Lo! the bright train their radiant wings unfold,
With silver fringed, and freckled o'er with gold;
On the gay bosom of some fragrant flower
They, idly fluttering, live their little hour,
Their life all pleasure, and their task all play,
All spring their age, and sunshine all their day."

MRS. BARBAULD.

"WHAT a pleasant life that must be to lead!"
methinks I hear my young readers exclaim;
"who would not join in the song,—

'I'd be a butterfly, born in a bower,
Where roses, and lilies, and violets meet?'

No tiresome books to bother the brains; no fagging at
lessons then; no cross looks, no angry words; no headaches, no stomach-aches, no whippings, no brimstone
and treacle; no anything but what is delightful and
pleasant; flitting about in the sunshine all day long.

1

and rocked to sleep at night in a lily bell, or some other such agreeable resting-place; sipping the sweet juices out of the flowers, and sporting in the air with companions that never get out of temper, and quarrel. Oh, that would be delightful! Yes, I'd be a Butterfly! Would not you?"

My dear young Master, or Miss, as the case may be, most assuredly I would *not* be a Butterfly. Nay, do not look so incredulous, but listen, and I will tell you why. In the first place I have no fancy to be snapped up by a winged monster two or three hundred times bigger than myself, as yon bright-winged flutterer has just been by the Swallow, that has a little hungry family up in the chimney there, and must find Butterflies or some other equally gay and thoughtless creatures wherewith to satisfy their wants. In the next place, I should not like to undergo such a series of changes and transformations as the Butterfly does, before he comes out in his beautiful silken dress, to live his little life of a few hours in the sunshine; and for all that is said in praise of " a short life and a merry one," by the thoughtless and careless among mankind, yet would I rather, if it so pleased God, live a long life, that I might have time to cultivate and exercise these high and noble faculties of the mind, which distinguish man from the rest of creation, and so exercise them as at once to glorify my Maker, and benefit my fellow-creatures. Nay, nay, my young friends, do not wish to be a Butterfly, nor any other merely *soulless thing ;* you have within you an immortal principle—

"A vital spark of heavenly flame,"

as the poet has finely termed the soul, which the Butterfly has not; which the most sagacious and long-lived of animals has not; for the salvation of this soul of yours a great price has been paid, a tremendous sacrifice offered, and young as you may be, I would have you think seriously of this. You are not a Butterfly, thank God that you are not! Never wish to be one! Do not lead a Butterfly kind of life, as too many do, flitting and fluttering, and sporting away the precious time given you for other purposes. Be diligent, be useful. Headaches and heart-aches, too, you must have, and many hard lessons you must learn, even when your schooldays are over; for it is ordered by an all-wise Providence, that the human soul shall be purified by trouble and affliction, and so prepared for the better land towards which we are all journeying. The end of the Butterfly is here; your end is in eternity. Think of that, and think, too, of the many pleasures which you enjoy, of which the Butterfly can know nothing; intellectual pleasures—pleasures of thought and feeling; warm affections and lively hopes are yours, out-gushing from your own heart and bosom, and from the hearts and bosoms of those to whom you are deár, and watching round you like angels wherever you move. You can speak, and write, and think, and above all you can *pray*, and *be prayed for*. Here is a privilege! For the poor soulless Butterfly there are none of these good things.—

> " Its little hour of sunshine o'er,
>   It passes from the view,
>   To breathe the breath of life no more—
>   It is not so with you.
>   Your soul shall from the tomb arise
>   In beautiful array,
>   To dwell for aye in Paradise,
>   And everlasting day."

WHAT IS A BUTTERFLY?

"Who can follow Nature's pencil here?
    Their wings with azure green and purple glossed,
    Studded with coloured eyes, with gems embossed;
    Inlaid with pearl, and marked with various stains
    Of lively crimson through their dusky veins."

MRS. BARBAULD.

HAT is a Butterfly?—An insect. True; and the name we are told is a literal translation of the old Saxon word *Buttor-fleoze*, applied to those silken-winged flies, because they usually become plentiful in the butter season. I have next to ask you what you understand by an Insect?—A little crawling, or flying thing, with —— Nay, that will not do at all. Let us find out Johnson's definition of the word. Ah, here it is, in Latin *Insectum*, that which is cut—"Insects may be considered together as one great tribe of animals: they are called Insects from a separation in the middle of their bodies, whereby they are cut into two parts, which are joined together by a small ligature, as we see in wasps and common flies." You have no doubt noticed this remarkable peculiarity of the Insects here named; it is especially conspicuous in the wasp, the lower part of whose yellow body looks as

if it would drop off at every motion. We have heard
a very slim and genteel lady spoken of as having a
waist like a wasp, but hope she was not waspish in
other respects. Another meaning for the word insect,
given by the great dictionary-maker, is "Anything
small or contemptible." Let us illustrate this meaning.
—" Sir," said a little upstart man, desirous of impress-
ing the person he addressed with a due sense of his
consequence, "do you know what sect I belong to?"
" I should say, by the look of you," was the good-
humoured, yet cutting reply, " to that called Insect."

We must not, however, consider that because things
or persons are small, that they are *therefore* mean and
contemptible; arrogance and undue assumption of im-
portance always make people so; but in the world of
Nature we find so much that is wonderful in design,
and beautiful in construction, in the minutest creatures,
that to the philosophic mind they can never be so. With
the poet Cowper,—

> " In the vast and in the minute we see
> The unambiguous footsteps of the God
> Who gives its lustre to the insect's wing,
> And wheels His throne upon the rolling worlds."

I am now going to introduce to you another member
of the learned family of OLOGIES. In a former volume
of this series you made the acquaintance of two or

three members of this family.* This is rather a tall individual.—What do you think of him?

ENTOMOLOGY.

FIVE syllables, thus—En-to-mol-o-gy. Let us see what account Dr. Johnson gives of him. None at all! Not in the big folio? Nay; then he must have sprung into existence since the great lexicographer's time; for all that he stands upon Greek legs—*entom*, an insect; and *logos*, a discourse. Now you know what it means literally—a discourse on an insect, or as generally applied, "That part of ZOOLOGY, or Natural History, which treats of Insects." By this science we are conducted into the most extensive and populous province in the whole empire of nature, and shown a greater diversity of form and colouring, and more surprising adaptations of means, than is presented to us by any other branch of physical science. Truly has it been said that "Entomology claims it as its right to demonstrate the existence and perfections of that Almighty Power which produced and governs the universe. It is one chapter in the history of creation, and naturally leads every intelligent mind to the Creator; for there are no proofs of His existence more level to the appre-

---

* 'Nests and Eggs of Familiar Birds.'

hension of all than those which this chapter offers to the understanding."

> " In an insect or a flower,
> Such microscopic proofs of skill and power
> As hid from ages past God now displays,"

says the poet Southey, in allusion to the wonders revealed by the microscope in the natural world, and especially in that branch of it with which Entomology has to do.

But it is to one particular division of the insect tribes that I have now to direct the attention of my readers.

LEPIDOPTERA

is the name given to that order of insects in which " Beautiful Butterflies " are included. Here is another long word not to be found in Johnson's dictionary— Lep-i-dop-te-ra, five syllables, derived from two Greek words—*lepis*, a scale; and *pteron*, a wing. Butterflies and Moths, then, are Lepidopterous or scaly-winged insects; if you observe one of them closely, you will find that the wings are covered with a fine downy substance like meal. Examined under a microscope this will be found to consist of minute scales of uniform size and shape, that is, upon one species of fly, for they differ considerably in this respect in different species, as the following drawings will show.

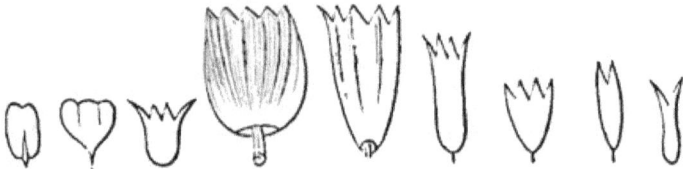

They are fixed to the wing by means of a fine pedicle, or stalk, similar to that of a plant, only so small as not to be seen by the naked eye. It is in these scales that the beautiful colours, which make the wings look like painted velvet, exist; if you rub them off, nothing but a thin transparent membrane remains; this is veined all over, much like the skeleton leaves which you may have seen, and these veins no doubt answer the double purpose of canals for conveying nourishment to the frame, and of ribs for giving it mechanical strength. A naturalist named Leiuwenhoek has counted as many as four hundred thousand scales upon the wings of the Silk Moth, and some of our British Butterflies are four times as large as this; there are foreign Moths which sometimes measure nearly a foot across the wings: think of the number of scales required to cover them. It has been said that "a modern Mosaic picture may contain eight hundred and seventy Tesserulæ, or separate pieces, in one square inch of surface; but the same extent of a Butterfly's wing may sometimes consist of no fewer than one hundred thousand, seven hundred and thirty-six: he would be a rich man indeed who had as many guineas. How long would it take him to count them, suppose he were to pick them up sixty a minute, and work ten hours a day at that rate? There's an exercise in mental arithmetic for you.

Now let us go back to our subject, which you know is Lepidopterous Insects—recollect that long word—or Butterflies, we were going to say, but the thought occurred that all these scaly-winged insects are not Butterflies; some of them are Moths, and some Hawk Moths. The three great divisions into which naturalists have divided this order of insects are, as you must try and remember, first, Butterflies, or *Diurnal*, that is, Day, Lepidoptera; second, Moths, or *Nocturnal*, that is, Night, Lepidoptera; and third, Hawk Moths, or *Crepuscular*, that is, Twilight, Lepidoptera; these names indicate their different seasons of flight. It is with the first division only that we have to do at present; and this forming a genus of itself, is distinguished by a generic name, and here it is—

PAPILIO.

Look at it well, now, so that you may know it again when you see it, as you often will in books of Natural

History. *Pa-pil-io,* pronounced *pa-pil-yo ;* it comes from the Latin, and means a Butterfly, which is all I can tell you about it. In botanical works you will sometimes see plants spoken of that have *papilionaceous* flowers, that is, with petals something in shape like the wings of a Butterfly, as the sweet pea has, and several other beautiful ornaments of the garden, with which you must be familiar. This resemblance of the bright-hued flowers to the Butterflies' wings have been often alluded to by the poets, one of whom, named Thomas Moore, describes

> " A child at play,
> Among the rosy wild flowers singing,
>   As rosy and as wild as they,
> Chasing with eager hands and eyes
> The beautiful blue Butterfles
> That fluttered round the jasmine stems,
> Like winged flowers or flying gems."

In another part of the same poem, which is called " Lalla Rookh," the name of an Eastern princess, we find a scene described in which

> " Sparkle such rainbow Butterflies,
>   That one might fancy the rich flowers
>   That round them in the sun lay sighing,
>   Had been by magic all set flying."

But let us get on with our lesson. Butterflies, then, we have learned, are a day-flying *genus,* called *Papilio,* of the Lepidopterous or mealy-winged order of that class of living creatures called Insects, the study of whose nature and habits is termed Entomology. The great Swedish naturalist, Linnæus, arranged all the flying, walking, creeping, and swimming things known in his time into six *classes;* these classes included several *orders;* the orders various *genera;* and the genera distinct *species;* more or less numerous as the case might be. This was called the Linnæan system of natural history. I need not explain to you the principles on which it was based, nor tell you in what respects it differs from the systems of the illustrious Frenchman Cuvier and other naturalists. One of these days I may perhaps do this, but at present it is scarcely necessary to puzzle your brains about it. I want you clearly to understand what a Butterfly is; to learn one letter of the great alphabet of nature first, and it will assist you in acquiring the rest. Now, what is a Butterfly? "An insect of the Lepidopterous order —— " Nay, you are going to tell me what it is *called* merely; undoubtedly it is all that, but it is also something more—a wonderful manifestation of the wisdom and the goodness of the Almighty Creator, as I am now going to show you.

HISTORY

OF

THE BUTTERFLY.

E will now trace the history of the butterfly, from the time it was a tiny egg, not so big as the head of a good-sized pin, glued by the mother insect to that particular kind of leaf on which the caterpillar, that it will shortly turn to, feeds. Now, here is a wonder at once; the Butterfly, recollect, does not feed upon leaves, but the sweet juices extracted from flowers and those by no means flowers of the plant from which it derives nourishment while in the caterpillar or *larva* state, as it is called. How then should it know the particular description of food suitable for its crawling progeny, in every respect so unlike itself? We can only say that God teaches it. INSTINCT is the name generally given to the mysterious knowledge which seems to direct all the members of the brute creation. Man, you know, has Reason for his guidance, animals have not; still they are guided, and often more surely to the desired end, than man with all his boasted reason The poet Pope has said

"Reason raise o'er Instinct as you can,
In this 'tis God directs, in that 'tis man."

Well, then, guided by this mysterious principle called Instinct, the Butterfly affixes its eggs just where the young Caterpillars, when they issue forth, are sure to find plenty of suitable food ready for them.

To the naked eye all Butterflies' eggs look round, and pretty much alike, but by the following representation of those of six different species, magnified, you will see that they are by no means so. You cannot fail to be struck with the beautiful regularity of the shapes and the markings. Do they not seem to say to you—

The hand that made us is Divine!

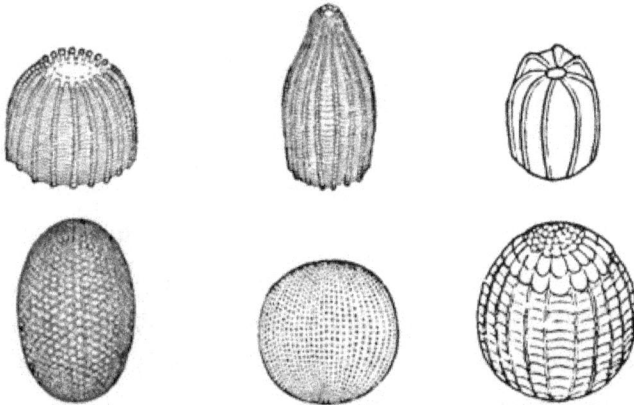

The term CATERPILLAR is supposed to be derived from two old French words—*acat*, food or provision, whence comes the term cates, sometimes used by English authors; and *piller*—to rob or plunder, the origin of the word pillage. There is here an obvious reference to the voracious habits of these creatures, the most destructive of any to vegetation. In Scripture they are spoken of as eating up what the other insect ravagers have left, as you will see by turning to the fourth verse of the first chapter of the prophet Joel. In Hebrew the Caterpillar is called a *consumer*, and well does the cultivator of the land know it to be such; it begins to eat directly it comes out of the egg, and continues to eat, eat, nothing but eat, except grow, which it does very fast, and crawl from place to place in search of fresh food, of which it sometimes consumes more than double its own weight in twenty-four hours. Think if you were to do this, what bakers', butchers', and grocers' bills your parents would have to pay.

The body of the caterpillar is, as you know, long, and nearly cylindrical, that is, like a tube, or pipe; it is divided throughout into twelve *segments*, as they are called, that is, divisions, as though pieces of thread were tied round it at equal distances, and drawn sufficiently tight to make slight indentations. The skin, which covers this body, is usually soft and *membranous*, that is, web-like, covered with little lines, which cross and re-cross each other, as in a piece of network; sometimes, however, it is of a *coriaceous* texture, that is, tough, like leather; in both cases it is very *flexible*, so that the creature can easily turn and twist itself about.

Most of the *Larvæ* of the Diurnal Lepidoptera have sixteen legs, six of which, that is, three pairs, are placed on the three first segments of the body, the part which corresponds with what is called the *thorax* of the winged insect; these foremost legs appear to be the ones principally used for locomotion or travelling; they are of a hard substance, rather wide where they join the body, and gradually tapering down to the bottom, where they terminate in a strong claw, by which they can draw themselves along. Like the leg

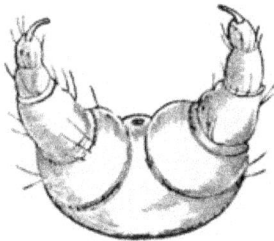

of the future fly, each of them is divided into several segments, or, in other words, it has several joints, as represented in this figure. You will see that he is rather a bandy-legged fellow—this Caterpillar, but there is wisdom displayed in this, as in every other part of his structure, how admirably are these legs adapted for clasping and holding fast.

Now for the other, or *pro-legs*, as they are sometimes called; *pro*, in Latin, means for; therefore this is as much as to say these are not legs, but substitutes for legs, and such it appears they really are; their principal use seems to be to support the body, to the hinder part of which they are attached; a pair on each segment up to the ninth, and the remaining pair on the last; by adhering to the twigs or shoots, on which the animal crawls. You stare at the word animal applied to a Caterpillar, but it is quite correct. All living creatures are animals; you are one, and I am one, and the invisible animalcule that sports in a drop of water is one. But we will not stop to discuss the point now, having the pro-legs of our crawler to examine. I have said that they are soft and fleshy, or membranous legs, or things that in some measure answer the purpose of legs. They are of a cone, or sugar-loaf shape, and can be lengthened and shortened at pleasure, like the horns of a snail; each of them is terminated by a triangular-shaped foot, if foot it may be called, at the bottom of which is a flat surface, or what may be termed the sole; on the inner edge of this is a row of small hooks, or claws, consisting of a long and short one alternately placed. When the foot is extended, these claws are turned outwards, and their curved points find inequalities on which they can take hold on almost any substance, however smooth it may appear. Here are two cuts;—one exhibiting this curious pro-leg with the foot expanded, and the other showing how a pair of them embrace and hold fast to a twig or branch.

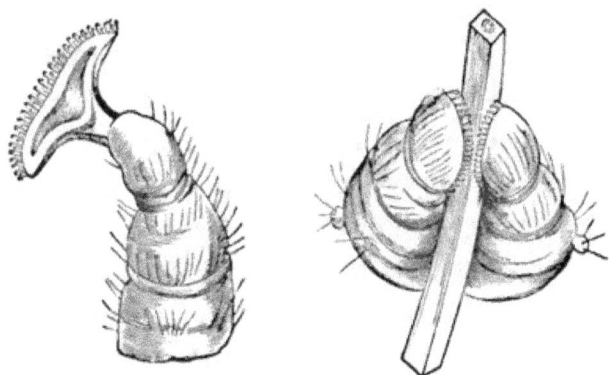

We have now to speak of the head of the Cater-pillar, which is harder than the rest of the body, and is generally composed of two oval plates joined toge-ther; in the next cut is represented the under side of this head; that part marked *b* is the mouth, consisting of an upper lip, with a notch in the centre; *c c* are the

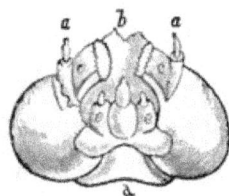

two strong mandibles, or jaws; and *d* is the under lip, near the top of which is a cone-shaped protuberance, from the centre of which, through a small hole, issues the fine silken thread which serves several important purposes in the changes which the insect undergoes; the *spinnaret* is the name which has been given to this organ; on each side of it is a similar shaped, but smaller, protuberance, marked *e e;* these are generally thought to be the *palpi*, or feelers.

The rapidity with which a Caterpillar eats proves that its cutting machinery is in good order, and well arranged for facility of working; if you watch the creature feeding, you will observe that the leaf on which it intends to operate is taken edgeways, and held steady between two of the fore feet. Before commencing, the body is stretched out as far as possible; the rapidly-moving jaws take off piece after piece, which is instantly swallowed, and at every bite the head is drawn in, until it comes close home to the legs, which hold the leaf, when another extension takes place, and another series of bites is commenced, till, by and by, the substance bitten is hollowed out like a half-moon. The notch in the upper lip, being even with the place where the jaws unite, appears to answer the purpose of a groove, to keep the margin of the leaf steady, and guide it in the proper direction.

The eyes and the *antennæ*, or horns, have now to be noticed; the first appear to the naked eye like two little dark spots; the microscope shows that each of them consists of six distinct eyes, or lenses, arranged in a circle; they are placed in the fore part of the upper side of the head, and are therefore not seen in the cut. The *antennæ*, or horns, although usually large in the perfect insect, are small in the Lepidopterous larvæ, being composed of two or three short joints, fitting into each other, like the tubes of a telescope, like which, too, they can with some species be drawn in, so as to be almost imperceptible; the letters *a a* point out their situation in the cut.

You will have noticed that Caterpillars differ very much in appearance;—some being smooth, and others rough and hairy; with some again the hairs are long and silky, with others short and harsh; some have them in tufts placed at regular intervals along the back or sides of the body; indeed they are seldom, if ever, irregularly placed; but whether few or many, a certain order of arrangement may be observable in the little tubercles, or lumps, from which they spring; these tubercles are placed in rows across the middle of the segments, and each segment is armed—for the spines may be considered as defensive armour—with a transverse, that is, a crosswise, series, varying in number from four to eight. The mode of arrangement and structure of these spines is exhibited in the section here given of a not uncommon species of British Caterpillar.

Why some Caterpillars should be hairy and others

smooth it is impossible for us to tell; in His wisdom God has so ordered it, we may be sure, for some good and useful end. Upon the hairy species, it has been

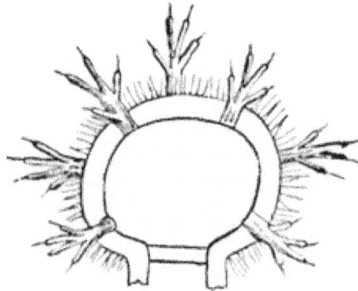

observed that the birds seldom prey ; and there is no doubt some sufficient reason why they should be especially guarded from their enemies.

Many of the Caterpillars are very beautifully coloured and marked, appearing to have silken coats, embroidered with gems; this is more especially the case with the larvæ of some of the foreign day Lepidoptera ; in this country those of the crepuscular division, or Hawk Moths, are the most beautiful, and to these we shall have to refer in another volume.

There is nothing more curious in the whole history of the Caterpillar than its *moultings*, or changes of skin, of which there are, at least, three, before the full size is attained. When it gets too big for its coat, the creature by a laborious effort, which is a painful, and at times even a fatal one, wriggles itself out, and leaves the old garment for whoever may take a fancy to it; but there is no occasion to go and get measured for another, as there would be with you or I, for a new one quickly forms around the body, fitting as nicely as need be. Sometimes the colours of this new coat are different from those of the old one, and the markings have a fresh arrangement, as if the fashions of the season had been studied in its preparation; but no, it was made in accordance with certain laws ordained from the beginning, in which there is no variableness nor shadow of turning. The fashion of a man's coat, or a boy's jacket, changes constantly; but the fashion of a Caterpillar's skin never: under certain circumstances it is always the same; its variations are regular, so that one who has observed these matters can tell what kind of covering a particular species at a particular age will have. The Naturalist will know at once the nature, and the character, and the age of a certain smooth or hairy worm which crosses his path, or feeds upon the leaf near at hand, by the shape and markings of its outer garment; suppose we could tell as much of men and boys by the cut and colour of their coats and jackets! Would not this be convenient?

Much might be said about the internal structure of the Caterpillar, but we cannot say it now; the same wisdom of design and nicety of adaptation is exhibited there as in the parts more exposed to observation.

We will now suppose the creature passed through all the changes incidental to the *larva* state;—this word in Latin means a mask; in this state the future fly is hidden or masked; *Larvæ*, the word we have

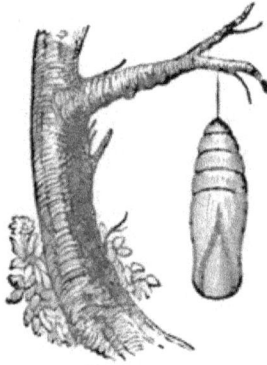

hitherto used, is the plural—it has eaten its own weight over and over again, may be a hundred times or more, in nettle, or cabbage, or some other leaves, and has now arrived at a ripe old Caterpillar age, so it prepares to go into what is called the *pupa* state, the second stage of its strange existence. *Pupa* in Latin means a puppet, a baby, a thing wrapped up, swaddled, as they say, of no particular shape—a lump; such is a *chrysalis*, as the insect, in the *pupa* state, is called. Singular, *Chrysalis;* plural, *Chrysalides* from the Greek *chrisos,* golden. The Latin authors termed these *aure-*

lia, from *aurum,* gold; because the cases of some of the
pupa have a glittering appearance, as though they were
partly composed of that metal. You may have seen
one, that of the Tortoise-shell Butterfly, generally found
near a bed of nettles, on which the Caterpillar feeds;
it is rugged, uneven, angular in shape, as is also that
of another common species, the Cabbage Butterfly;
this latter is of a greenish-yellow colour, marked with
black spots.

Some Caterpillars, when about to change into chrysalides, creep into holes and crevices, others bury themselves in the earth; some envelope themselves in a
tissue of fine silk, as the silkworm does; this is called
a *Cocoon,* of the origin of which word we must confess
our ignorance; my readers will do well to try and find
it out. Some of these chrysalides, again, suspend
themselves to a bough, or other convenient object, by
means of a silken cord, sometimes fastened round the
middle, sometimes to one end, of the oblong case, as
shown in this cut, designed by Thomas Hood to illustrate the popular ballad " I 'd be a Butterfly." Here
the part downwards is the head. A very interesting
account is given by a naturalist named Reaumar of
the way in which the Caterpillar proceeds to effect this
suspension, and change from the *larva* skin into the
pupa case; but it is too long for quotation here; by-
and-by you will, I hope, read it for youself in some

larger Entomological work. It is related that it sometimes takes as long as twenty-four hours to effect these changes. Reaumar observes that " it is impossible not to wonder that an insect which executes them but once in its life should execute them so well. We must necessarily conclude that it has been instructed by a Great Master: for he who has rendered it necessary for the insect to undergo this change, has likewise given it all the requisite means of accomplishing it in safety."

When first the larva skin is thrown off, the chrysalis is soft and tender; it is covered with a sticky kind of fluid, through which portions of the future Butterfly can be clearly seen; like a thing closely packed up and put away until it is wanted. Gradually the fluid covering hardens, and closely enveloping all the parts of the insect, forms a case impervious to wet or any other atmospheric influence.

> "It was a shrivelled shrouded form,
> Though but of late a living worm;
> A caterpillar it had been,
> Once clad in suit of silken green;
> But now how changed by nature's laws!
> Where are the eyes, the legs, the jaws?
> No signs of being can one trace
> In the cold mass; its outer case,
> Like cere cloth round a mummy spread,
> Is passive, motionless, and dead."

And there it swings, or lies wrapped up in its silken

cocoon, or hid in the earth, or some snug hole or corner, until the time arrives for the release of the little prisoner, when—

> "Lo! the shrouded thing
> Loosed from its earthly covering,
> From shape uncouth and dusky hue
> Like some fair vision springs to view,
> A glossy wing in burnished pride
> Unfolding rises from each side;
> Its tapering form in beauty dressed,
> Like gold dust o'er a yellow vest;
> Whilst hands unseen had giv'n the power
> To gather sweets and suck the flower,
> It is a Butterfly, as bright
> As ever sparkled in the light."

And this third state of its existence is called tne *Imago*, or perfect state of the insect; this term is the Latin for Image, whence comes Imagination, that faculty of the mind which produces images of things unknown, &c.

I will now recall the stages through which the insect has passed before it assumed that glorious shape—that complete *image* of beauty, in which it now hovers and flutters before our eyes—

> " The brightest and the lightest thing
> That flits about on sportive wing;"

and to impress these more deeply on your memory,

will place, beneath the name of each, its pictorial representative.

The appearance of these creatures in their various states of Caterpillar, Chrysalis, and Butterfly, is so strikingly dissimilar, that it was long a general belief

LARVA.   PUPA.   IMAGO,

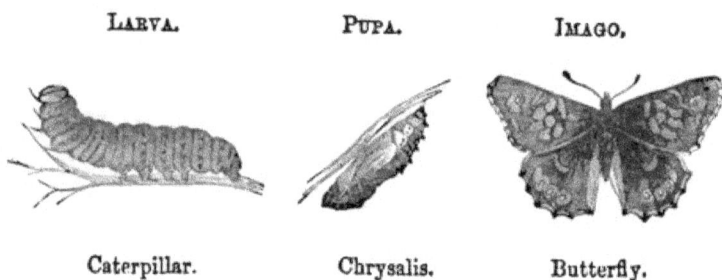

Caterpillar.   Chrysalis.   Butterfly.

that they underwent, at each successive stage, a complete transmutation, or change from one being to another; "but it is now clearly seen," says the naturalist Swammerdam, " that within the skin of the Caterpillar a perfect and real Butterfly is hidden, and therefore the skin of the Caterpillar must be considered only as an outer garment, containing in it parts belonging to the nature of a Butterfly, which have grown under its defence by slow degrees, in like manner as other sensitive bodies increase by accretion," that is, by growing or gathering of new matter. In every Caterpillar, therefore, it would seem that from the earliest period of its life there exists the germ or seed, if I may so call it, of the future fly, even as in the interior of the unexpanded bud may be discovered the germ of the future flower. So that all these changes are but a series of developments, a throwing off of outer coats, so to speak; and showing more and more clearly the perfect form within. All this is very wonderful, and worthy of our closest study and attention. So striking did these changes which the insect undergoes appear to the ancient Greeks, that they regarded the Butterfly as the fittest emblem of the soul of man; accordingly we find that their word *Psyche*, pronounced *Syke*, with an accent on the last letter, signifies both the human soul and a Butterfly. And very beautiful and appropriate is this emblem. How like a rising from the tomb and soaring upward to the realms of light, is this bursting forth of the imprisoned fly from its dark chrysalis chamber, to spread its glittering wings and float aloft in the golden sunshine. The poet Rogers has written some fine lines upon this subject, which, I think, you will be able to understand and appreciate: here they are—

> "Child of the sun! pursue thy rapturous flight,
> Mingling with her thou lovest in fields of light;
> And, where the flowers of paradise unfold,
> Quaff fragrant nectar from their cups of gold,
> There shall thy wings, rich as an evening sky,
> Expand and shut in silent ecstacy.

*   *   *   *

Yet wert thou once a worm, a thing that crept
On the bare earth, then wrought a tomb, and slept;
And such is man! soon from his cell of clay,
To burst a seraph in the blaze of day."

And now let me read to you a lesson of moral instruction, which the natural history of this insect is
calculated to enforce; I will do it in the words of a
German fabulist, or writer of fables; this story is called
an Allegory:—A humming-bird met a Butterfly, and
being pleased with the beauty of its person and glory
of its wings, made an offer of perpetual friendship.
"I cannot think of it," was the reply, "as you once
spurned me, and called me a drawling dolt." "Impossible," exclaimed the humming-bird, "I always enter-
tained the highest respect for such beautiful creatures
as you." "Perhaps you do now," said the other,
"but when you insulted me, I was a caterpillar. So
let me give you this piece of advice: never insult the
humble, as they may one day become your superiors."
No, never insult the humble, nor despise that which is
mean-looking, merely because it is so. The smallest
and lowliest creatures have in them much that is
worthy of admiration, aye, even of respect, for they are
all the works of one Great Creator, and you know not
what they may be destined to.

"Look round creation, and survey
Life springing forth from life's decay :

In gladsome April view the tree
Resume its verdant livery ;
From bars of ice the river freed,
Pursue its course along the mead :

And earth, escaping from th' embrace
Of winter, show a joyous face.
E'en thus the worm, though lowly found,
Groping its way upon the gronnd,
May yet revive, a creature fair,
And wander 'midst the realms of air."

We will now examine a little into the structure of
the perfect insect—a common Butterfly ; and a truly
wonderful piece of living machinery we shall find it.
The body is divided into two principal divisions, called
the *Thorax,* meaning the breast or chest, and the
Abdomen, or stomach. The first of these is composed
of three segments, something like those described on
the Caterpillar, and the last of six or seven ; these two
sections of the body have, as in most insects, a very
narrow line of connection ; they are both covered with
hairs, more or less long in different species. The
upper one is always thicker than the lower, because to
it are attached the organs of motion—the legs and
wings, the muscles of which, especially those of the
latter, are numerous and powerful. The legs are six
in number, and this is the case with all true insects:
there are three on each side, and their points of junc-
tion with the body are pretty close to, and at about

equal distances from, each other. In each leg there are three principal divisions, the thigh, the *tibia*, or pipe, and the *tarsi*, or toes, just as there are in the human leg, only the shape differs considerably, as you know. A Butterfly would not look well with silk stockings, because he has no calf; the thigh is often fringed with long hairs, and the *tibia* armed with a spur in the middle, and two others at the top; the *tarsi* are five-jointed, and furnished with two claws at the extremity, which are often what is called *bifed*, that is, cleft in two; this word comes from the Latin *bi*—two, whence also comes *biped*—a two-legged creature, and several other words beginning in the same way. When I tell my little boy he is a *biped*, he says, "and you are another," and he is quite right, for we are all *bipeds* —*bi*, two; *pede*, foot—literally, two-footed, as *quadruped* is four-footed; and *tripod*, a kind of stool that stands upon three feet; from the Latin *quatuor* four, *tres* three; also the roots of many English words. But why do I trouble you with all this about roots and derivations? because I am desirous of making you look into things, to be thoughtful and inquiring, so that you may know the *reason* why this or that name is applied to a certain object or operation. This is the way to learn. Some Butterflies are called *tetrapod*, or four-footed, because, although they have six legs, they have but four feet, two of the legs being what are termed *spurious* or false, there being only one joint in the *tarsis*, and consequently no claw or foot.

From the legs we go naturally to the wings, they being situated very close together ; of these butterflies have two pairs, the upper pair being generally of a triangular form, and the lower pair nearly circular; the shape, however, as well as the size, varies greatly in different species; and the difference in the colours and markings is, as you know, great indeed. I have already spoken of the mealy powder or scales (feathers they are sometimes called), in which the colour resides, and may, therefore, pass on to observe that the thin membrane, of which the wing itself consists, is traversed by small tubes called *nervures,* filled with air and air-like fluid, which acts so as to expand the wing when the fly first issues from the *pupa* state, and to strengthen and keep it extended afterwards ; the principal *nervures* rise from the point where the wing joins the body, and where there is a horny scale covered with tufts of hair, so that it resembles a little epaulet, such as the soldiers wear; and the parts between the main pipes, as they may be called, are termed by naturalists *areolets,* that is, little spaces, from *area,* an open or flat surface between lines or boundaries.

Naturalists consider the upper wing of the Butterfly divisible into three of these spaces, and across them from branch to branch ramify, that is, spread out, smaller pipes, so as to form a complete network, as I have before observed. In some species these minute veins are much more numerous than in others, and in some they scarcely or at all occur.

It may be observed that Butterflies, when at rest, generally carry their wings upright, so that the backs of them meet together, and only the under sides are visible; and here is an obvious mark of distinction from the Moths, which keep theirs flat, even with the ground, or whatever they may rest on, and so show only the upper part; the flight of the Butterfly, too, is more wavering than that of the Moth generally. It does not go in a direct line, but keeps rising and falling, and herein often lies its safety from the pursuing bird, with whose beak it is level one instant, and the next far above or below it. Thus it is that the weaker creatures are often able to elude or baffle the pursuit of the stronger by superior cunning or agility—a something which is given as a compensation for want of power, and a means of preservation from their foes.

Not all the Butterflies, however, are distinguished by this irregular and fitful mode of flight; some of the larger sorts, which are strong upon the wing, go more directly to their object, and sometimes even make way against currents of air which would make the less weighty and powerful flutterers seek shelter in a lily-bell, or a tent of green leaves. A practical entomolo-

gist can generally tell the species of a Butterfly by its
mode of flight, so much difference is there in this
respect.

One very important part of the structure of the insect
I have not yet described, that is, the head, the most
conspicuous and curiously constructed organ of which is
the long flexible tube used for sucking the juices out of
the nectaries of the flowers ; this is of a *cartilaginous* or
gristly substance, arranged in rings, woven together, as
it were, by means of minute fibres, so that it can be
curved or twisted in any direction, with great ease and
rapidity. This tube is, in reality, the mouth of the
creature, for it has no other, nor any occasion for one,
as it lives by suction, and is hence classed among the
Haustellated (from the Latin *Haustus,* a draught) or
suctorial tribes of insects. The *Proboscis* or trunk of
the Butterfly is divided into two distinct portions, which
can be separated throughout their whole length, each
portion being grooved on the inner side ; they form,
when united, a sort of canal of a squarish shape, through
which passes the nourishment which the insect sucks
up. The union of the two halves of the trunk is effected
by the interlacing of an immense number of threads,
which form a kind of fringe along the edges, and so
close is it that the canal is perfectly air-tight ; and on
each side of it there runs from top to bottom a circular
passage, the use of which does not appear to be very

clearly known, although it is thought most likely to be the transmission of air for the purposes of respiration or breathing, which, however, is mainly carried on by means of pores, or very minute openings in the sides of the body, called *stigmata*, from *stigma*, a mark, these pores presenting the appearance, under the microscope, of little pits or dots. Near the outer extremity of the trunk, which, when not in use, is curled up like the spring of a watch, are generally a number of projections resembling leaflets; a scientific naturalist would call them *papillæ*, the Latin for nipples. Reaumur supposes that the use of these is to steady the organ, by adhering to the sides of the flowers into which they are inserted. In order that my readers may have a clearer understanding of the structure of this wonderful little instrument, I have caused three cuts to be executed. *A* is a mag-

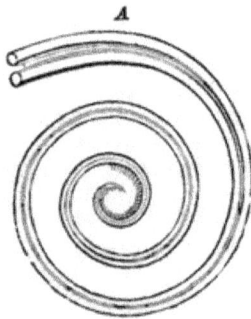

A

nified view of the trunk, exhibiting its general form, and the *papillæ* near the tip. *B* is a highly magnified section in which the mode of connection is shown, with

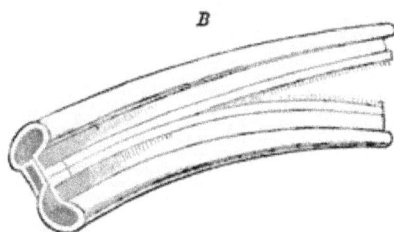

the central and two lateral or side canals.   *C* is another

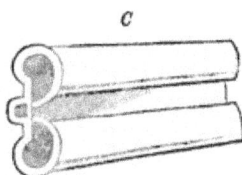

section showing the under side.   Is it not marvellous
that all this is in a little tube not much thicker than a
hair; human skill and ingenuity can do much, but they
would surely be at fault here.

We have seen that the proboscis of the Butterfly is
composed of two distinct parts, which may be considered
as occupying the same place in the general structure of
the insect, as the *maxillæ* or under jaws of the *man-
dibulated* tribes.   We have here two Latin words, and
when I tell you their meaning, you will understand
what I wish to express; *maxilla*, then, means the
upper jaw-bone—the termination *æ* makes it plural;
mandibula is the Latin for jaw, so the term above used

would signify tribes that have jaws, as much the larger number of living creatures have. It was, perhaps, not quite correct of me to say that the Butterfly had no mouth, for there is a little cavity just below the insertion of the trunk, which, although it hardly deserves the name of a mouth, and does not appear to be used at all as yours and mine, or even that of the Caterpillar is, yet it must, I fancy, be so called.

And we must now speak of the *oral* appendages, that is, the parts which belong to the mouth. The word comes from the Latin *os* or *oris*. Now, if you should hear or read that a person has or intends to communicate with another *orally*, you will understand that it means by word of mouth—that is, by speech. This little cavity that I spoke of is covered by a small triangular plate, " which must be regarded," says an authority on these matters, " as the *labium* or under lip." There is another Latin word for you to remember. Those letters of the English alphabet, such as b, p, v, f, m, which are pronounced chiefly by the lips, are called *labials*, you know, or should know. You have, no doubt, noticed two short points projecting from the front of the head of a Butterfly—I do not mean the horns, which are long, slender, and nobbed at the ends —but I mean what are called the *labial palpi*.

In describing the head of the Caterpillar, I spoke of the *palpi* or feelers; this latter term comes from the

Latin *palpus*, and it expresses the action of feeling gently or timidly, just as the snail does, you know, which draws in its horns directly they come in contact with anything. The term also means to flutter, hence you may hear a weak person say that his heart *palpitates*, when he has used great exertion, or been over-excited. Now, both these meanings will apply to the *palpi* of the Butterfly, which sometimes has a quivering or fluttering motion, and I have been thus particular in explaining them, in order that you may see what great significance there is in scientific terms generally. Ignorant people often wonder what use there can be in these long names, but you here see that they have a use, and you may also see how necessary to a right understanding of them is a knowledge of the Greek and Latin languages, on which they are chiefly founded. These languages may be considered as the keys to all the sciences—by all means try and obtain possession of these keys, for they will well repay you for years of study; with them to guide you, and the habits of observation and method, and analysis, which their very acquisition will induce, much that now seems difficult and strange will be clear and simple, and full of order. Now let us go back to the *palpi*, it would be *palpa* if only one, but that there are two is quite *palpable*, that is, plain—they can be seen and felt, and those who have examined them through a microscope, tell us that they

are three-jointed, and in shape like this.   In this cut,
the letter *a* indicates the organ I have been describing;
and *b* the *antenna,* plural *antennæ.*   The term I have

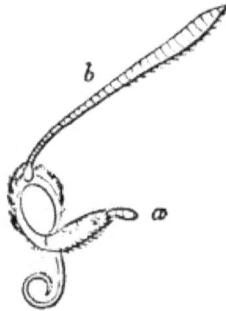

used before in describing the insect in its *larva* state;
but I did not then speak of its derivation.   The Latin
word *antenna* means the sail-yard, as it is called, of a
ship, that is, the piece which crosses the mast, and to
which the canvas is affixed; these horns of the insect,
in some cases sticking out in nearly opposite directions
to each other, suggested the idea of the cross-pieces on
a mast, and so the same Latin name was applied to
them.   They are generally of considerable length, and
consist of a great number of joints, usually increasing
in thickness towards the extremity, where they form a
kind of knob; this formation you must often have
noticed; it is peculiar to the Diurnal Lepidoptera, and
by its variations of shape affords an obvious distinction

between different genera. In the Moths the *antennæ* are often beautifully feathered and fringed, but we do not find this the case with the Butterflies.

I have now only to speak of the eyes, which are of two kinds, the *stemmetic* or simple eyes, which are usually two in number, and placed on the crown of the head, where, if seen at all, which is not the case in all species, they appear as little clear spots, nearly covered with hairs and scales. Their use as organs is somewhat questionable, but they may be of service in some way, as yet undiscovered by the investigation of scientific men. The ordinary or compound eyes, as they are called, are plain enough; they occupy a great part of the head on each side, and project from it like a half globe; unlike the human eye, which can move in various directions, they are fixed, but to compensate for this they are composed of an immense number of little lenses, each of which is capable of reflecting a perfect image, so that on whatever side you approach a Butterfly (and it is the same with most other insects), it can see you without moving its head or rolling its eye-ball in the socket, as you or I would be obliged to do, to be aware of any danger approaching from a direction other than the front. Naturalists tell us that they have counted as many as seventeen thousand three hundred and twenty-five lenses in a singe eye; double that for the pair, and it gives thirty-four thousand six hundred

and fifty *little circular looking-glasses*, to a single pair
of eyes, contrived and fashioned for an insect destined
to live but for a few hours. I can tell you nothing
more wonderful than this, so I think I may as well
bring my description of a Butterfly to a conclusion, the
more especially as I have mentioned all with which it
is necessary for you to be acquainted about the history
and structure

> Of that wonder of wonders, the bright-winged fly,
> That flits to and fro in the azure sky ;
> That has died, and been buried, and sprung from the tomb,
> To live amid sunshine, and beauty and bloom.

As the subject of my present volume is ' British
Butterflies,' I would say a few words with especial
reference to them. They are neither so large nor so
beautiful as some of the foreign ones, yet are there
many of the native species remarkable for elegance of
form and richness of colour, as the examples here given
will serve to show ; and, by the way, here let me mention one peculiarity in the wings of the Butterfly,
which should not go unnoticed—the under, as well as
the upper, side of this wing, is beautifully painted by
the Divine Hand, often of a different pattern from that
displayed on the outer surface. How elaborately
finished are all God's works! How perfect in every
part !

The Lepidopterous order of insects ranks next to the

Coleoptera, or Beetles, in point of numbers; in England, where the variable and moist climate is certainly
unfavorable to their increase, there are not much less
than two thousand species: between two and three
thousand of the Diurnal Lepidoptera alone have been
discovered and described all over the world; and of
this division of the order seventy-five species inhabit
Britain. At the end of the volume will be found
a complete list of these, arranged according to their
genera, with a specification of their times of appearance, and of the places they usually frequent.

BUTTERFLY HUNTING.

BUTTERFLY HUNTING.

IN a boy the desire to capture one of these beautiful insects that comes dancing and fluttering before his eyes, as if to invite pursuit, is natural and instinctive; off goes the cap without any other thought than how the prize is to be obtained; trampled flower borders, torn trousers, even bruised limbs, are as nothing in the account, and probably he sits down at last, flushed, heated, tired, and disappointed; but it is only to start up again and renew the chase, should the same chance of a capture offer itself.

> " Before your sight
> Mounts on the breeze the Butterfly, and soars,
> Small creature as she is, from earth's bright bowers
> Into the dewy clouds."

And while you stand wondering what has become of the insect which seemed but now within your grasp, another brighter and more beautiful still issues from the variegated tulip cup, and as Mrs. Hemans has it,

" like an embodied breeze at play," wavers about amid the flowers; and off you go again in hot pursuit, like the child of Vigillia, in Shakspere's play of 'Coriolanus,' but not like him, I hope, to get into a rage and destroy wilfully the poor fly, because you have a tumble or two in endeavouring to catch it. " I saw him," says Valeria to the boy's fond mother, " run after a gilded Butterfly, and when he caught it, he let it go again, and after it again, and over and over he comes, and up again—catched it again; or whether his fall enraged him, or how 't was, he did so set his teeth and tear it; O, I warrant how he mammocked it." This is by no means a pleasing picture, but I fear it is too often realised among young Butterfly hunters, who, if they do not get into a passion and designedly destroy the object of their pursuit, frequently do so accidentally in their efforts to secure it. So delicate and fragile is the creature, that but the brush of a cap, or the slightest pressure of a finger, will, if it crush not that wonderfully organized frame, and render it lifeless, take away much of its beauty, and with it, no doubt, much of its capacity for enjoyment. And after all, if their efforts are crowned with success, how poor is the prize gained. The poet Byron has some beautiful lines on this subject which I should like you to read:—

" As rising on its purple wing
    The insect queen of Eastern spring,
    O'er emerald meadows of Kashmeer,
    Invites the young pursuer near;
    And leads him on from flower to flower,
    A weary chase, and wasted hour,
    Then leaves him as it soars on high,
    With panting heart and tearful eye."

The poet thus, after saying, which is true, how full-grown children, that is, men, are lured by things which, if as beautiful, are also as fragile as Butterflies, supposes that the prize is gained, and goes on—

" The lovely toy so fiercely sought,
    Hath lost its charm by being caught,
    For every touch that woo'd its stay,
    Hath brush'd its brightest hues away,
    Till charm, and hue, and beauty gone,
    'Tis left to fly or fall alone."

We have heard of an enthusiastic entomologist who followed a Butterfly for nine miles, in the hope of capturing it; and this must be set down to the account of ardour in scientific investigation. Not all grown Butterfly hunters, however, would we hold so excused, many, very many, more unfortunate insects are swept down with the net, and pinned out in the collecting case, than are required for the purposes of science; and this wholesale destruction of insect life we think scarcely compatible with that abhorrence of cruelty, and rever-

ence for the works of the Great Creator, which is
enjoined by Him

> " Who formed the gilded fly, and o'er its wing,
>   A picture, decked in rainbow colours, drew;
>   To sport amid the sunshine of the spring."

My young readers must not suppose from this that I
would harshly condemn in them that love of Butterfly
hunting which is common to all children. I have felt
it myself, and know how strong is the temptation to
follow "the painted toy."   There is perhaps no prettier
sight than such as that described by Grahame, in his
poem on July, when

> "At noontide hour from school the little throng
>   Rush gaily sporting o'er the enamelled mead,
>   Some strive to catch the bloom-perched Butterfly;
>   And if they miss his mealy wings, the flower
>   From which he flies the disappointment soothes."

In gazing on such a scene as this, I become a
boy again, and am half inclined to join in chase
myself, and to whoop and halloo with the maddest
there; but then the thought occurs — would it be
right to risk even doing injury to a creature so
wonderfully fashioned, and to shorten the little span
of its joyous existence. Let this thought restrain
your hand; run and leap as much as you like, it is

natural to your age and good for your health to do so, and follow

"The little fly with wings of sunbeams,"

from end to end of the green meadow, but do not attempt to catch it; let it be to you as a sacred thing, sent by God to beautify the earth, and delight your eyes, but by no means to be wantonly injured or destroyed.

I should like to repeat to you many beautiful poems which have been written on the Butterfly, and many striking observations that have been made on it by naturalists, such as Messrs. Kirby and Spence, whose admirable work on Entomology I hope you will read when you are older, but I cannot do so now, as my book is but a small one, and I shall want all the available space to describe the species of which figures are given. In the poems above spoken of allusion is sometimes made to the Butterfly as a fop—a light careless thing; thus Thompson says—

"While a gay insect in his summer shine,
The fop, light fluttering, spreads his mealy wings."

The same poet too, in his 'Castle of Indolence,' speaks of it as an emblem of pleasure :—

" Behold ! ye pilgrims of the earth, behold !
　See, all but man with unearned pleasure gay ;
　See her bright robes the *Butterfly* unfold,
　Broke from her wintry tomb in prime of day,
　What youthful bride can equal her array ?
　Who can with her for easy pleasure vie ?
　From mead to mead with gentle wings to stray,
　From flower to flower on balmy gales to fly,
　Is all she has to do beneath the radiant sky."

Leaving unquoted those fanciful lines by Roscoe—
' The Butterfly's Ball,' and ' The Butterfly's Funeral,'
with which many of my readers must be familiar, as
well as other Butterfly poems which I have at hand, I
will conclude this Introductory Chapter with some
lines by the quaker poet, Bernard Barton, who thus
addresses the bright insect—

　　BEAUTIFUL creature, I have been
　　　Moments uncounted watching thee,
　　Now flitting round the foliage green
　　　Of yonder dark, embowering tree ;
　　And now again in frolic glee,
　　　Hov'ring around those opening flowers,
　　Happy as nature's child should be,
　　　Born to enjoy her loveliest bowers.

　　And I have gazed upon thy flight,
　　　Till feelings I can scarce define,
　　Awakened by so fair a sight,
　　　With desultory thoughts combine

Not to induce me to repine,
  Or envy thee thy happiness;
But from a lot so bright as thine
  To borrow musings born to bless.

For unto him whose spirit reads
  Creation with a Christian eye,
Each happy living creature pleads
  The cause of Him who reigns on high;
Who spann'd the earth, and arch'd the sky,
  Gave life to everything that lives,
And still delighteth to supply
  With happiness the life He gives.

This truth may boast but little worth,
  Enforc'd by rhet'ric's frigid powers:
But when it has, its quiet birth
  In contemplation's silent hours;
When summer's brightly peopled bowers
  Bring *home* its teachings *to the heart*;
When birds and insects, shrubs and flowers,
  Its touching eloquence impart.

Though many a flower that sweetly deck'd
  Life's early path, but bloom'd to fade:
Though sorrow, poverty, neglect—
  Now seem to wrap their souls in shade;
Let those look upward, undismay'd,
  Turn thorny paths in anguish trod,
To regions where in light array'd,
  Still dwells their Saviour and their God.

Sport on then, lovely summer fly,
  With whom began my votive strain :—
Yet purer joys *their* hopes supply,
  Who by Faith s alchemy obtain
Comfort in sorrow, bliss in pain,
  Freedom in bondage, light in gloom,
Through early losses, heavenly gain,
  And life IMMORTAL through the TOMB.

DESCRIPTION OF SPECIES.

1. Swallow-tail.   2. Marbled White.

SWALLOW-TAIL.

PLATE I.—FIG. I.

PAPILIO MACHAON is the name applied by Linnæus and most naturalists to this large and beautiful species of Butterfly; the meaning of the first, or generic name, has already been sufficiently explained; the second, or specific name, has reference, it is likely, to the peculiar shape of the two lower wings of the insect, from each of which issues, as will be seen by a reference to the plate, a projection shaped like a dagger or knife, the Latin word for which is *Machera*. There are only two British Butterflies which have wings of this singular form; and it is likely that few of my readers have seen either of them, except, perhaps, in the cases of the British Museum, or some other entomological collection. The species to which our attention is now directed, is the largest Butterfly found in Britain, sometimes measuring as much as three inches and three quarters across the expanded wings, of which it will be seen the ground-colour is yellow, with black markings; these markings being remarkably bold and distinct, and the broadest of them being powdered over with very small dots,

which look like gold dust; and some of the yellow
portions have a like sprinkling of little blue dots; on
each hind wing, as it is called, is a patch of a rich
purple colour, and on the inner side a red spot like a
fiery eye. The body of the insect is like black velvet
powdered with gold; and altogether it is so richly
dressed, so gracefully formed, and so large of size, as
well to deserve the title bestowed upon it by one
naturalist, named De Geer, of *Papilio Regina*, the
Queen Butterfly.   And where think you does her
Queenship delight to hold her court?   On the sunny
uplands, and in dry pastures, gay with Nature's gold
and silver—the buttercups and daisies? Nay, amid
the moist fens and reedy islands, which abound in the
shires of Lincoln, Cambridge and Huntingdon; there
it is that the Swallow-tail Butterfly is most plentiful,
and may be taken from the beginning of May to the
end of August, by those who are venturous enough to
follow it into the swampy grounds, amid which it
delights to dwell.   Here, too, may be found the beautifully-marked green and yellow caterpillar, with its
black bands, and rows of little rings like eyelet holes
across the back, feeding away upon the marsh parsley,
the wild carrot, and the fennel, and other plants which
flourish in such moist situations.   From June to September is the feasting time; in the latter month the
change into a chrysalis takes place, and a curiously-

shaped case it is that the voracious feeder goes to sleep in—all points and angles, of a light green colour, with yellow marks, and a row of dusky dots down each side.

There are several English counties besides those before named in which this Butterfly has been found, but in none of them is it so plentiful as in those low and moist districts,—

> Where the dragon-flies dart 'mid the rustling reeds,
> And the great sleek water-rat builds and breeds;
> Where the moorhen glides through the waving sedge,
> And leads her young to the marshes' edge;
> Where the stagnant pool is with duckweed green,
> And gnats rise in clouds when the air's serene;
> And the alder grey like a sentry stands
> To warn men's feet from the swampy lands.

Since the draining of the fens in Lincolnshire, and other parts of the country, it is said that the beautiful Swallow-tail has become less plentiful, and a fear has been expressed that entomologists may in time lose this most conspicuous ornament of their cabinets. Well, better so than suffer such waste and barren spots to remain uncultivated; such nurseries of fever and pestilence to send forth their unhealthy exhalations, and spread sickness and death throughout the surrounding districts. Wholesome food and pure air for man is of far more consequence than the preservation of an insect, however beautiful and rare; so let them drain Whittlesea Mere, as they have lately been doing, and turn the fens into farms, where something more than flocks of cackling geese can be bred and fed; even although the race of Swallow-tails should become extinct in consequence. But we have little fear of this; for many centuries to come there will be marshy waste places, where umbelliferous plants, such as the wild parsley, carrot, and fennel grow and flourish, and where, if anywhere, one may expect to find this Queen of British Butterflies, in company, perhaps, but this would be a rare chance indeed, with the SCARCE SWALLOW-TAIL, called by naturalists *Papilio Podalirius,* of which but a few specimens have been taken in England. Why it was called *Podalirius,* we cannot tell,—this was the name of a son of the ancient Greek physician, Esculapius. Both in shape and markings this insect differs considerably from the more common kind—the wings are more pointed and slender; there are no red eyes in the lower ones, nor patches of purple, but dark blue spots shine on each side in the place of them. The Caterpillar is shorter and stouter, green with yellow and red markings. It is said to feed on the leaves of the apple, sloe, plum, peach, and almond; the chrysalis is light brown.

MARBLED WHITE.

PLATE I.—FIG. II.

PAPILIO, or *Hipparchia Galathea*, is the Latin name of this species. You will remember what I told you about the *genera* into which naturalists have divided the members of the animal and vegetable kingdoms. *Hipparchia* is a *genus* of Butterflies including many species, of which this is the largest, and perhaps also the handsomest; most of the others differ from it greatly in the colours of the wings, which are brown of various shades, sometimes tinged with red, and sometimes with dusky yellow. The Caterpillars are nearly always green, more or less marked with yellow, and sometimes with red. They feed on different kinds of grasses; that called the cat's-tail grass being the favourite food of the larva of the Marbled White, which obtains its English name from having its wings chequered and veined as we sometimes see marble; these wings, however, are not, strictly speaking, white, but a pale yellow; in some there is a much greater proportion of black than in others.

But we have got something more to say about the scientific title. *Hipparchus*, we find, was the name of an ancient astronomer; but why it applied to this genus of insects we cannot tell. A group of crustaceous animals, or shell-fish, has had the term *Galathea* given to it by some naturalists, but why, our reading does not inform us, any more than it does wherefore a certain species of Butterfly should be so called. *Lac* is the Latin, and *gala* the Greek, for milk; hence we have *galaxy*, that collection of stars in the heavens called the milky way; *Galactites*, precious stones, of a milk-white colour, etc. The wings of this beautiful fly are of a milky, or rather of a creamy tint, and hence perhaps may have originated the specific name of the insect, which appears in its perfect state in the months of June and July.

It may be found in every county in England, although only in certain spots, or localities, as they are called; usually in lonely places, in and near woods, or on wide open downs or moors. It measures from one to two inches and a half across the extended wings; the Caterpillar is about an inch and a quarter long, with yellow stripes running along the sides of its thick, green body; the chrysalis is scarcely half the length, of a light brown colour; it is usually found suspended from the stalk of the cat's-tail, or some other grass.

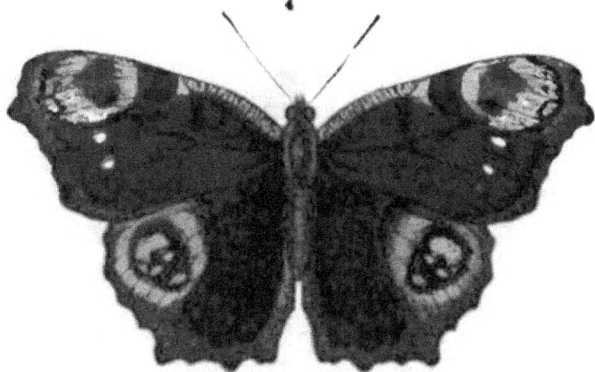

1. Scotch Argus.   2. White Admiral.   3. Red Admiral.

4. Peacock Butterfly.

SCOTCH ARGUS.

PLATE II.—FIG. I.

IPPARCHIA BLANDINA. This is a ve..
beautiful, although at first sight it appears a
sober-coloured Butterfly; the wings, of a rich
brown, ornamented with veins and patches of bright
copper colour, and beset with small eye-like dots, are
well deserving of a close examination. They measure,
when extended, from an inch and a half to two inches
across, and their outline is at once bold and graceful;
the under sides of the hinder pair, seen when the insect
is at rest, have on them two broadish waves of a grey
tint, and both pairs on either side are margined with
yellowish brown, which being crossed with darker
lines, looks like a delicate fringe.

This insect, which is not uncommon in many parts
of Scotland, especially the southern counties, has also
been captured in Northumberland, Durham, and Yorkshire, as well as some other parts of England, in which
country, however, it is by no means plentiful.

The Caterpillar is light green, with brown and white
stripes, running lengthwise, these would be termed
longitudinal stripes; the head is of a reddish colour;

what it feeds on does not appear to be known; nor as far as we can learn, has the chrysalis been discovered.

Respecting the name, we can only say that Argus refers to the little spots on the wings, that being the designation in the ancient mythology of one who is said to have had a hundred eyes; so we say of a very watchful person, like your schoolmaster, for example, that he is argus-eyed—he sees everything. But for that matter, *all* Butterflies are Arguses, for, as I have already shown, they have each a great many more than a hundred little reflecting lenses; to these real eyes, however, the term does not refer, but to the eyelike rings with dots in the centre, which appear in the wings of this and many other species. The Wood Argus (*Hipparchia Ægeria*) and the Arran Argus (*H. Ligea*) are two of these; the latter closely resembles the one we have just described. It is very rare in Britain, having only been found in the Scottish Isle of Arran, whence it takes its name. The former is an elegant fly, with dark brown wings, specked with buff-colour, from which the little eyes peep out; the under sides are much lighter, beautifully streaked and mottled. This insect is found in woods and lanes all through the country, from April to August. It is sometimes called the Wood Lady, or Speckled Wood Butterfly.

We have before mentioned three generic names—

Blandina, *Ægeria*, and *Ligea*, to what do *they* refer? The nearest approach we can make to the first, in the Latin, is *Blandusia*—the name of a fountain near the country seat of the ancient Roman poet, Horace. Bland, you know, in our language, means soft, smooth; and this comes from a Latin root having the same sound and signification.

Ægeria, in the old mythology, was the name of a disconsolate maiden, who took to crying so, that it was thought best to make her a fountain at once; and so travellers tell of the grotto of Ægeria, near Naples, whose trickling waters do not taste at all like salt tears.

Ligea was the name of one of the Nereides; and who are they? you may well ask. Oh, they were nymphs of the sea, fifty of them, and all sisters—a nice little family; they lived in grottoes and caves on the sea-shore, all decked with shells and shining spars, and when they went out for an airing, they rode upon dolphins. They had the power of calming the waters, and so sailors in old times used to pray to them for prosperous voyages. But what has all this to do with Butterflies? I am sure I do not know; ask the grave naturalists who gave such pagan out-of-the-way names to our little brown Arguses, and led us away into the region of myths and fables.

WHITE ADMIRAL.

PLATE II.—FIG. II.

APILIO, or *Limenitis Camilla.* This is an insect most usually found in the shady woodlands of Essex, Sussex, Hampshire, Suffolk, Kent, and Middlesex; it is by no means a common species in any part of England, and is remarkable rather for grace and elegance of form than richness of colour. The wings, which, when expanded, sometimes measure as much as two inches and a half across, are, on the upper side, of a dull brownish black, with a broad band crossing each; this is very irregular in shape, and varies considerably in breadth in different individuals; there are also white specks scattered here and there about the wings, the under sides of which are of a light reddish brown, with the white markings of the upper sides showing through; near the body of the insect, they become of a grey tint; the back and upper parts of the body also assimilate to the outer surface of the wings in hue.

The Caterpillar, which feeds on the honeysuckle, is from an inch to two inches long; of a green colour, ringed and ornamented with faint red lines, and tufts of hair along the back. The angular-shaped chrysalis is pale brown, marked with black lines and golden spots. It has a curious projection, like a knob, on the back, and where the head of the insect is are two distinct lumps. It is found suspended by the point to a honeysuckle or some other stem.

English naturalists, we believe, have not been fortunate enough to discover this gilded mummy case, although many a hunt has taken place in the hope of doing so; neither has the insect in the larva state been found in this country. Those few who have been fortunate enough to have opportunities of observing the fly on the wing, have been struck with the peculiar grace of its movements. Thus the Rev. Mr. Haworth says, "The graceful elegance displayed by this charming species when sailing on the wing, is perhaps greater than can be found in any other we have in Britain." And the Rev. Everett Sheppard tells us that "In its beautiful flight, when it skims aloft, it rivals the Purple Emperor, which it strongly resembles in appearance. It seems, however (unlike the latter), to avoid the sunbeams; for it frequents the glades of woods, where it rapidly insinuates itself by the most beautiful evolutions and placid flight through the tall underwood, on each side of the glades, appearing and disappearing like so many little fairies," and seeming, as the rev. gentleman might have added, to say to the pursuer,

"Through the wood, through the wood, follow and find me,
Search every hollow, and dingle, and dell."

With respect to the scientific name, it may be observed that *Limenitis* may have reference to the flashing appearance of the white bands on the wings, as the insect flies about in the dusky glade; *leme* being an old English word, signifying a ray or flash. *Camilla*, in the mythology, was the name of a queen of the Volschi, so swift of foot, that she could pass over a field of corn without bending the ears, and over the sea without wetting her feet. We should rather believe that our dusky-winged *Camilla* could do this than the Volschian queen, however swift-footed she might have been.

This insect, we are told, is not rare on the continent of Europe, where there are four other species belonging to the same genus, and closely resembling it.

RED ADMIRAL.

PLATE II.—FIG. III.

APILIO, or *Vanessa Atalanta.* This large and magnificent Butterfly is by no means uncommon. It comes out of the chrysalis in August, and may frequently be seen flitting from flower to flower in the garden, seeming to prefer the large globe-like dahlia and the little green blossoms of the ivy, two very different objects of regard; doubtless, however, the nectar of the one is as sweet as that of the other, and little need the gaily-dressed insect care about richness of colour, for he has that himself; look at him now, as he sails proudly in the autumn sunshine, which seems to sink into the soft velvety down of his rich brown wings, and give them a golden gloss.

The colour is deep, approaching to a black, and across the middle of each upper wing a painter seems to have drawn his brush filled with the brightest vermilion; along the bottom edge of the lower pair, there is also a band of the same colour dotted with dark brown or black; there are several irregular white specks on the upper ones towards the angles of their widest expansion, where they measure in some instances as much as three inches across; near the tips is an indistinct wave of purple, which catches the eye in certain lights, and all round the margins of both pairs is a fringe of white arranged like a chain of little crescents.

The inner sides of the wings can hardly be described; they are beautifully mottled and variegated with black, brown, buff, steely blue, and other metallic tints; the ground colour is lighter than that on the outer side, the red and white markings of which gleam through like flashes of different coloured flame.

Oh, a glorious fellow is the Red Admiral! and so bold and fearless, too, as an admiral should be; you may come close up to him as he sits sunning himself upon the flowers, and examine him at your leisure; but do not attempt to touch him, unless you really want a good specimen for your cabinet; and even in that case, it is better, perhaps, not to risk the spoiling of these magnificent wings, but to search for a caterpillar: they are not difficult to find among the nettles where they feed. The colour is dusky green, with a yellow line running along each side; it is sparingly covered with short hairs, and its length varies from an inch and a quarter to one and three quarters; it often conceals itself by drawing together several of the nettle leaves by means of silken threads.

When you have captured it, give it some of the seeded tops of the plants, for this is the part which it seems to prefer. By and by it will turn into a chrysalis of a greyish brown colour, having golden spots sprinkled over it; the back is rough and ridgy, the tail end sharply pointed, and the opposite end bearing a remarkable resemblance to the snout and head of a miniature pig.

It is generally at the latter end of August or beginning of September that this beautiful fly makes its appearance, in its new suit all fresh and glossy, and it generally disappears by the end of October; a few, it seems, manage to get into some snug hole or crevice, and live through the winter, coming out again for a short time in the spring, but with wings sadly rubbed and faded—insects which " have seen better days." Usually, however, the life of the gallant Admiral, is, as short as it is, no doubt, merry.

But why is it called *Vanessa?* why *Atalanta?* Let us see if we can find it out. The first you know is the generic name, and the genus to which it is applied, includes some of the most common, as well as the most beautiful of the British species; several of them we shall presently describe. Our business is now with the Red Admiral, or Alderman, as he is sometimes called, because, perhaps, aldermen once wore, and in some places still do wear, scarlet gowns edged with brown fur, making a mixture of colours something like the wings of this noble fly. But what about *Vanessa?* well, here again we are at fault; we can find no name, proper or common, no Greek or Latin root, at all like it, except *vanesco*—to vanish—in the Latin tongue, and therefore must give it up.

Atalanta I could tell a long story about, did I wish to entertain my young readers with old fables; she, too, according to mythological story, was one of the swift-footed, and did some very wonderful things, how many hundred years ago I am afraid to say; and now, that her name may not be clean forgotten, which I think it might just as well be, naturalists have bestowed it upon this gorgeous insect, although it appears quite a misnomer, as the Admiral is rather a heavy flyer for a Butterfly, and the English name which he bears is that of a gentleman, and not a lady.

All the flies of the *Vanessa* genus are remarkable for their robust bodies, and the thick texture of their wings; they are strong and handsome, rather than light and elegant; that is, comparatively speaking, for lightness and elegance are inseparable from the whole race of diurnal Lepidoptera.

PEACOCK BUTTERFLY.

PLATE II.—FIG. IV.

APILIO, or *Vanessa Io*. We have here another magnificent species, on which many of my readers must have often looked with intense admiration, for it is common enough throughout the greater part of England, especially in the more southern counties; it gets rare towards the north, and in Scotland is seldom found. Old writers called this Butterfly *Omnium regina*, that is the Latin for Queen of all, and we ourselves are half inclined to rank it before all others for beauty and richness of colouring.

It is scarcely so large as the Red Admiral, seldom measuring more than two inches and a half across the expanded wings, which are of a rich dark brownish red, the upper pair several shades lighter than the lower; at places there is an inclination towards a purple shade, and a large eye, or, as naturalists call it, an *ocellus*, from the Latin word *oculus*—an eye, covers a considerable portion of each of the four wings; being situated, in the upper pair, near the extremities farthest from the body, and in the lower, also near the tips and outer margins ; these *ocelli* are much like the many-coloured spots on the tail of the peacock, and hence the insect has been named after the bird.

Io was the name of a heathen goddess, about whom there is told a cock-and-a-bull story, not worth repeating, more especially as we cannot learn that it has any relation to the habits or appearance of the insect to which the name is applied, and the under side of whose wings are very different from the outer, being of a uniform dark brown colour, traversed by black waving lines, which form a complete network, through which, like stars through a mist, the blue and white spots on the outer sides faintly peep.   The body of this insect is blackish, clothed with rust-coloured hairs, and the legs are of a dull yellow colour.

The Caterpillar is also black, or nearly so, with numerous white points dispersed in rows across the body, which is partially covered with hairs; the legs are a rusty red. Its habits are gregarious, that is, numbers live and feed together; it feeds on two very common species of stinging nettle, amid which it may be found without much difficulty about July.

The chrysalis is, both in shape and colour, much like that of the Red Admiral; there is something more of a green tinge in it, and the indentations along the back are deeper. The resemblance of the thicker end to the head of the swine is also more marked and distinct, and piggy here appears to be adorned with two stick-up ears, which is not quite in accordance with swinish fashion.

LARGE TORTOISE-SHELL.

PLATE III.—FIG. I.

APILIO, or *Vanessa Polychloros.*   The Tortoise-shell Butterfly is one of the commonest
of our native species, but the Great Tortoise-
shell, or Elm Butterfly, as it is sometimes called, must
not be confounded with this; it is larger, handsomer,
and also very much rarer; in the colours and markings
of the two species there is not a great deal of difference; neither is there in the shape of their wings; the
smaller is perhaps the more graceful of the two; the
larger the more bold and noble—more rounded and full,
and sturdy, so to speak, as though one were the per-
sonification—bless us, that is a long word—well, then,
the figure, the showing forth of manly beauty, and the
other of womanly loveliness.   If we saw the two flying
together, we might take them to be brother and sister,
so great is the family likeness, and yet they are quite
different species, although belonging to the same genus;
and a naturalist would never take one for the other,
however closely they might approach in size, as they
sometimes do.   Oh, these are sharp fellows, these
naturalists; we sometimes fancy that they have micro-

1. Large Tortoise-shell.   2. Camberwell Beauty.

scopic eyes, that is, that they carry microscopes in their
heads as well as in their pockets; they can detect such
minute points of difference, and so quickly too. They
can see things that nobody else can see, and they tell
such wonderful stories about the creatures that inhabit
drops of water, and grains of sand, that one is at times
inclined to think they have only been dreaming, and
ought to begin their almost unbelievable stories with
the line of the poet Coleridge—

"In a vision once I saw."

But then they tell us how we may convince ourselves of
the truth of their statements, and bid us look through
their glasses at some of the monsters of the invisible
world, which are frightful enough in pictures; but all
alive and kicking are, oh dear! enough to make a
nervous person go clean out of his senses—"off his
head," as they say in some parts of England.

One of these days I mean to write a little book for
your instruction, all about the animalcules, and the uses
of the microscope, and then we will get our engraver to
serve you up such a dish, or rather a number of plates,
of odd-shaped creatures, which inhabit the world out of
sight, that if you won't be startled and surprised, that's
a pity!

But what is our *Vanessa Polychloros* doing all this
time? enjoying himself in the sunshine no doubt,

like an idle fellow as he is, for the song says, you know that

> " The Butterfly was a gentleman,
>   Of no very good repute,
>   And he roamed in the sunshine all day long,
>   In his scarlet and purple suit."

There he goes fluttering, fluttering, with his gaily-painted wings spread out, with the tips as much perhaps as three inches apart, without appearing in the least incommoded by the long name he carries about with him, and which, if he should let it fall so as to break it into two halves, would turn out to be made up of the Greek words *Polus*—many, and *Chlorus*—spot, so that it means, literally, many-spotted; and this we must confess to be a very appropriate title, better than all the *Atalantas* and *Camillas*, and such like proper, or rather improper, names of fabulous persons. It was most likely on a bright July day that the Great Tortoise-shell Butterfly—you know why he is called *that*, because he is marked like the shell of a tortoise—burst from his chrysalis prison, and he will most likely leave this gay and busy scene before the end of September; either dying, as is generally the case, or stowing himself away in some nook or cranny, where his superb wings can be kept dry for a little spring exercise, should the weather prove at all genial; they get, however, sadly rubbed and rumpled, and cannot be made to look like new by all the dyers and scourers in Butterfly land. I shall not attempt to describe these wings, for all my readers have seen the Small Tortoise-shell fly, and often enough to know every marking; should any of them unhappily belong to that class of persons who go about the world with their eyes shut, they can open them for once, and look at the beautiful pictures with which our artist has illustrated this volume, and they will see at a glance what it would take many words to express in a very imperfect manner. The under side of the wings is a dull brown, approaching to a black, with here and there a streak of blue, and waving lines of dusky white, and irregular bands of deeper colour.

The Caterpillar of this species may be found most usually on the leaves of the elm tree, where they feed, numbers of them together, under little silken tents, spun out of their own bodies; their colour is a bluish-brown, and along each side runs a broad band of orange, which, with numerous tufts of yellowish hair, gives them quite a gay appearance. After the first change of skin they break up their family parties, and disperse to seek their livelihood singly; they are sometimes found on the willow and several kinds of fruit trees, especially the cherry.

The chrysalis is brownish flesh-colour, with golden

spots. It has the peculiar conformation at one end mentioned in the two preceding species.

This butterfly is to be met with in all the southern counties in England ; in some years it occurs in large numbers at particular places, but not so regularly as to be called a common fly anywhere in this country, where its near relative, the Small Tortoise-shell, is very abundant.

1

2

SMALL TORTOISE-SHELL.

PLATE IV.—FIG. I.

HIS is the *Papilio,* or *Vanessa urtica* of naturalists, and when I tell my readers that the
Latin for nettle is *urtica,* and that the Caterpillar feeds upon nettles, they will at once see the fitness
of the title. This Caterpillar is found in the beginning
of June, and again about the middle of August. In
the early stages of its growth it is gregarious like the
larva of the last species; the body is a dull greenish
brown, with paler lines down the back and sides; the
head black, as are also the tufts of hair; the chrysalis
is brownish, and shaped much like that of the larger
species. It is sometimes nearly covered with gilding,
and is generally suspended by the smaller end.

Everywhere in this country, and throughout the
whole of the summer, one sees the Little Tortoise-shell
Butterfly, of which species there appears to be two goal
deliveries in the year, one in June and the other in
September, so that by the time one batch of released
prisoners dies out another comes to take its place, and
thus the summer sunshine glanceth ever upon Tortoise-
shell wings, and frequently the spring sunshine too, for

LARVA, CHRYSALIS, AND INSECT OF THE SMALL TORTOISE-SHELL
BUTTERFLY. (*Vanessa urtica.*)

more of this species than of any other manage to preserve life through the rigors of the winter, and they
venture forth, when the blasting March gales have
sunk to rest for a time, and the sun has melted the
snow from off the hills, and the primrose blossoms and
the violet buds are just beginning to unfold their fragrant petals in woods and on mossy hedge-row banks.
But full often do they share the fate of the venturous
insects described by the Peasant Poet of Northamptonshire, John Clare, who says—

> " The Butterflies by eager hopes undone,
> Glad as a child, came out to meet the sun,
> Beneath the shadow of a sudden shower
> Are lost, nor see to-morrow's April flower."

The same poet, we remember, has another allusion
to the Butterfly making its appearance in March, much
to the delight of the old village dame; in this case,
however, it is one of the white species spoken of, most
likely the Common Cabbage Butterfly (*Pontia brassicæ*),
which sometimes comes out very early.

> " The old dame then oft stills her humming wheel—
> When the bright sunbeams through the windows steal
> And gleam upon her face, and dancing fall
> In diamond shadows on the pictur'd wall;
> While the White Butterfly, as in amaze,
> Will settle on the glossy glass to gaze—
> And, smiling, glad to see such things once more,
> Up she will get and totter to the door."

CAMBERWELL BEAUTY.

PLATE III.—FIG. II.

APILIO, or *Venessa antiopa.* This is the crowning glory of the British Butterfly collector's cabinet, and a happy man is he who gets a perfect specimen of an insect which is at once so rare and beautiful. It measures across the expanded wings from a little under to considerably over three inches, and is, therefore, one of the very largest of the native species, from all of which it differs greatly in shape, and still more in colour. Deep purplish-brown, of a rich velvety appearance, is the tint spread over nearly the whole upper surface of these wings; round the outer edges this colour deepens into a broad black band, which is ornamented with a row of violet blue spots, some of which are oblong, some of a crescent shape; attached to this band is what looks like a wide silken fringe, of a pale yellow or cream colour, slightly waved on the inner side, and sprinkled with little black dots at the angles; the top edges of the upper pair are slightly mottled with yellow, and on each, near the tip, are two spots of the same. Underneath the wings are a shining ashy brown, with a network of black waved lines all over them ; the yellow spots, so distinct on the outer side, showing faintly through.

The Caterpillar is large and black, with red spots along the back, and reddish legs; it is thinly furnished with clusters of hairs; it is gregarious, and feeds upon the leaves of the birch, willow, and poplar, and is said to be most usually found among the upper branches of these trees. The chrysalis is of a dull black colour, spotted with deep yellow, and of a very irregular shape. In neither the larva nor the pupa state has the insect been found, we believe, in this country, where its appearance occurs, except just here and there a single specimen or two, at long and uncertain intervals. About eighty years ago it was seen in great numbers in many parts of the kingdom, and again in 1819, but not since then, although almost every year one or more specimens are taken or seen.

This fly obtained its common name from having been first observed in the neighbourhood of Camberwell, in the county of Surrey; it is also sometimes called the Grand Surprise, and the White Border. The scientific name will take us again among the heathen gods and goddesses. *Antiope*, we are told, was a daughter of Nycteas, king of Thebes, who went through a variety of strange adventures, of the account of which we can only say, that if it is altogether false, so much the better; for, little as we like story-telling, what this *Antiope* is

said to have done we like far less.   The fair Camberwell
Beauty—the loveliest of British Butterflies—is by no
means complimented in having such a name bestowed
upon her.

We cannot refrain from quoting here a passage from
a monthly publication called 'The Naturalist,' wherein
this butterfly is very touchingly and pleasingly alluded
to.   Our young readers, we have no doubt, will
fully enter into the feelings and associations there
expressed :—

"It was on a summer evening, of early life, when
little more than a child, in rambling through a wood on
a holiday, my attention was drawn to a spray on which
rested a Camberwell Beauty.   I had never seen such
perfection before.   My eye rested on the rich dark
velvety wings, fringed with ermine white, relieved by
an inner border of metallic blue spots, like bracelets of
lapis lazuli.   At this moment I could mark the very
spot in the forest where this vision was revealed, and
well do I remember the thrill of delight with which I
captured and carried off my prize in triumph, to exhibit before a little knot of schoolfellows.   I can see
their uplifted hands, I can hear their exclamations of
surprise, as they beheld the splendid captive.   I can
recall their features and their forms as if now living,
though every individual among them has long since
been called away, and now possibly familiarised with

greater things than it is permitted man's philosophy to dream of here.

"But to me, trifling as this little incident may appear to many, the results through life have been neither unimportant, useless, nor uninfluential; for it is to it I stand indebted for many a happy hour. That 'poor insect' awakened a taste which has never slumbered; and the cultivation of Natural History has been my solace in times and seasons, when the mind required something to fall back upon, apart from the business and pursuits of the world. It so happened that from the time I have alluded to until a few summers ago, in one of the mountain passes of the Pyrenees, I had never met with a single living specimen of *Vanessa antiopa*, when, on a lovely day, on a spray the very counterpart of that of the days of my childhood, I saw the expanded wings of this insect, and the days of 'auld lang syne,' which first introduced it to my notice, came across my mind vivid and clear as though but of yesterday. This summer, again (and not unfrequently) I fell in with this associate of my early years. Children, indeed, may they be called of the sun. In the hot and sultry hours of noonday, they would flit by, rendering it almost impossible to watch their course; if in these flights two or three met in the glade, they paused in their speed, and fluttering together, so busied themselves in conflict of rivalry or affection, I know not which, that I more than once caught two at a time, and after admiring them, in gratitude for the benefit I had received at their hands, sent them forth once again to enjoy their summer revelries. At other times (I particularly recollect one occasion), in a wood on the summit of the Drachenfels, when the wind was rather keen, I found numbers resting on the backs of trees, in a state of stupor; they made no attempts to escape, and when thrown into the air their wings barely opened, or flapping feebly, eased their fall, or enabled them to seek repose on the stem of a rotten trunk."

1. Comma Butterfly.　2. Painted Lady.　3. Scarce Painted Lady.

COMMA BUTTERFLY.

PLATE V.—FIG. I.

APILIO, or *Vanessa C-album.* This curious Butterfly is distinguished from all other British species, by the deeply cut or indented outline of the wings; which, forming the kind of curve presented by the stop called a comma, gave occasion for the name by which it is commonly known. Naturalists we find term it *C-album,* the latter term meaning white; because there is marked on the under side of each of the lower wings a letter C in white, as distinctly on some individuals as though it had been painted in. This is a very singular distinctive mark, the like of which is to be found, I believe, on no other member of the Butterfly tribe.

The Comma is the smallest species of the *Vanessa* genus which inhabits Britain, the utmost extent of the wings seldom exceeding two inches; in colour it closely resembles the Tortoise-shell Fly, being reddish brown and black; there is, however, less variety in the shades and markings. The under sides of the wings are in some instances elegantly variegated with brown and grey of various shades, and metallic green; in others they are of a uniform dull brown, which held in different lights has a bronzy appearance. There are two broods of this fly in the year, one coming out in June, and the other in August or September; and it has been observed that the wings of the latter brood are much less bright and various in colour than those of the former.

The Caterpillar, which feeds on the elm, willow, honeysuckle, hop, nettle, and several other plants, is of a reddish brown, rather thickly set with hairs, and having a curious hairy projection on each side of the head, which is nearly heart-shaped. The chrysalis is a pale dirty brown, spotted with gold; shaped much like those previously described.

This is by no means a common fly in this country, that is generally speaking, for in some years it occurs in great plenty. York, Worcester, Dorset, Warwick, Suffolk, Gloucester, Herts, and Middlesex are the English shires where it has been chiefly taken. Also in Fifeshire, in Scotland, but in no other part of that country, except, perhaps, some of the southern counties.

A celebrated entomologist named Westwood, has observed a great variety in the shape of the wings of this insect, some not being nearly so deeply indented as others; so that there are Common Butterflies, the outlines of whose wings present rather the form of a note of interrogation, (?) as though they would bid the student in entomology to pause and ask himself, can it be a Comma or some other species; and so perhaps bring him to a semicolon, (;) or a colon, (:) or even a full stop. (.) While struck with the beauty of the insect, he will, it is likely, utter a note of admiration. (!)

PAINTED LADY.

PLATE V.—FIG. II.

APILIO, or *Cynthia cardui.* The genus
Cynthia into which we now pass, so closely resembles the *Vanessa* that it is considered by
some naturalists as only a sub-genus or kind of lesser
division thereof. *Cynthia* is a name sometimes applied
to the moon, as it was to one of the fabled goddesses
of the Grecian mythology, between whom and that
enlightener of the night, there was supposed to be
some mysterious kind of relationship. We have only
two *Cynthias* in Britain, and one of these can hardly
be called a native species, but a single specimen, I
believe, having been taken in this country ; of this I
shall presently speak. Now let us examine into the
merits of the Common Painted Lady, which is a truly
beautiful fly, graceful in form, and harmonious, if not
brilliant, in colouring. A reference to the plate will
show this better than any description, as far as the
general shape and outer side of the wings are concerned ;
but a chief beauty of this fly consists in the marking of

the under side, which looks like an exquisite piece of mosaic or inlaid work, in which the several pieces of red, orange, buff, olive, brown of various shades, black, and white, are nicely fitted in to form a diversified pattern, with small white veins dividing the compartments. Near the outer edge of the under pair is a row of four or five round spots, encircled with rings, so that they resemble eyes: on a close examination it will be seen that two of these are powdered in the centre with green, and two with blue. Between these and the edge of the wing is a row of small purple crescent-shaped spots. Over the whole of the under side of the upper pair is spread a delicate crimson flash, like that which tells the approaching dawn of a summer's day. The body of the insect is clothed with reddish brown hairs above, and white beneath.

The Caterpillar, which is found generally in July, is of a reddish brown colour, with interrupted yellow lines along the sides ; it is pretty thickly covered with hairs. It feeds upon thistles, nettles, mallows, artichokes, and several other plants, and lives singly. The chrysalis is light brown, with ash-coloured lines and golden spots; it is of an irregular angular shape.

The expanded wings of this insect sometimes measure two inches and three quarters across. It is generally somewhat scarce in England, although it occasionally occurs in great abundance ; in the year 1828 there passed over Switzerland a vast swarm of these flies, occupying several hours in the transit; two years previous to this the species was very abundant near London; it has been found in Ireland, and the more southerly of the Scottish counties.

SCARCE PAINTED LADY

PLATE V.—FIG. III.

PAPILIO, or *Cynthia Huntera*. Of this fly but one specimen is recorded to have been taken in Great Britain, and that was at Withybush, near Haverfordwest, South Wales, in the midsummer of 1828. It was thought to be but a variety of the common kind, but afterwards discovered to belong to a distinct species, occasionally plentiful in America, where it is said the Caterpillar is found about the end of the months of April and July, there being two broods in a year. It measures about two inches and a quarter across the expanse of the wings, which both in colour and shape are much like those of the *C. cardui*. We should have explained, by the way, that this word means a thistle.

The wild balsam and a kind of cudweed, called the obtuse-leaved, are said to be the food of this species in the larva state. Authorities differ as to the colour of its Caterpillar; its chrysalis we are told is " placed in the leaves of plants folded up and spun together," so says the Rev. F. O. Morris, in his beautiful work on ' British Butterflies.'

The markings on the under sides of the wings of this insect are yet more elegant and delicate than those of the last; there is also somewhat more grace and beauty in those of the upper surfaces, as will be seen by a reference to the plate.

PURPLE EMPEROR.

PLATE IV.—FIG. II.

APILIO, or *Apatura Iris*. Ever has purple been the imperial colour, that of the robes of kings and emperors, and altogether great persons, there is a fitness therefore in the name of this king of British Butterflies, than whose wings no colour more rich and magnificent ever decked the shoulders of the mightiest monarch upon earth. To see these wings in perfection, it must be in the broad sunshine, and rather at a side-angle of the vision than looking directly at them; in any other than a strong light they appear dull and lustreless—a rusty, dingy, brownish black, with just a tinge of purple, and no more; but even then it appears a noble fly—stout-bodied and strong-winged, with a flight like that of a bird, far up amid and above the topmost branches of the loftiest trees; there it flies, and there it sits on some projecting spray in right royal state, and where is the puny entomologist that with rod and net shall reach it down from thence? But sometimes it descends from these elevated positions to sail majestically amid the glades and along the wood-side hedge-rows, and then is the time of

7

danger to liberty and life.   Mr. Morris tells us of
one collector who stated that he took as many as
one hundred specimens in the county of Essex in a
fortnight!   What an arch regicide he must have
been.   Who ever heard of such a king-killer as this?
One hundred crowned heads in a fortnight!   Here
was a wardrobe! ten times ten sets of royal robes.
Bah! we do not like to think about it; so many beautiful insects deprived of life; so many bright flutterers
stopped in the mid career of their enjoyment, for what?
to satisfy a mania, as in too many instances the entomological furor may truly be called.   May be he did it
for profit; he dealt in Butterflies, and so depopulated
the glades of Emperors to increase the value of his
stock-in-trade.   Well, we scarcely know that any valid
objection can be offered to this; all we can say is, that
we hope no such necessity for obtaining a livelihood
will ever be laid upon us.   We would not have such an
amount of insect slaughter upon our consciences for all
the Butterflies in the British Museum, much as we
should delight in the possession of a good collection
of these most beautiful of the works of the Almighty
Creator.

The scientific name of this fly is a bit of puzzle.
Apatao, in Greek, means I cheat, trick, or beguile;
hence *Apaturia* was the name of a festival at Athens,
instituted in remembrance of a stratagem, by which

one king, who had challenged another to fight, put his opponent off his guard, and so killed him. One of the names of Jupiter we learn was *Apatenor*—the deceiver, and *Apate* is by one Greek author, applied to a plant. What connection, real or supposed, there is between all this and the generic name of our splendid Emperor we cannot imagine.

With the specific name *Iris* we have no such difficulty. In the mythology the messenger of the gods was called by this name, and the rainbow, more poetically than truthfully supposed to be her pathway from the regions above to earth, was also so termed; hence the word *Iris* came to signify that which was rich and various in colour, especially such as shifted and changed in different lights, or as we should say was iridescent. The rich purple flag-flower we call an iris, and so do naturalists term this glorious fly, whose dark wings are so richly overspread with purple down, and which is that well described by a British entomologist named Haworth.

" The Purple Emperor of the British oaks, is not undeservedly the greatest favourite of our English aureleans (Butterflies of which the chrysalis are marked with gold). In his manners likewise, as well as in the varying lustre of his purple plumes, he possesses the strongest claims to their particular attention. In the month of July, he wakes in the winged state, and invariably fixes his throne upon the summit of a lofty oak, from the outmost sprigs of which, in sunny days, he performs his aerial excursions; when the sun is at the meridian, his loftiest flights take place, and about four in the afternoon he assumes his station of repose. He ascends to a much greater elevation than any other insect, sometimes mounting higher than the eye can follow; especially if he happens to quarrel with another Emperor, the monarch of some neighbouring oak; they never meet without a battle, flying upwards all the while, and combating with each other as much as possible; after which they will frequently return again to the identical sprigs whence they ascended. The wings of this fine species are of a stronger texture than those of any other in Britain, and more calculated for that gay and powerful flight which is so much admired by entomologists.

" The females, like those of many other species, are very rarely seen on the wing; in three days I captured twenty-three (another regicide), nine of them in one day, and never took a female at all. The males are only to be taken with a long net, fixed at the end of a rod twenty or thirty feet long. There have been instances, though rare, of their settling on the ground near puddles of water, and being taken there. When the Purple Emperor is within reach, no fly is more easily taken; for he is so very bold and fearless, that he will not move from his settling-place until you quite push him off; you may even tip the end of his wings, and be suffered to strike him again."

Here is the character of his majesty given with great truth and freedom; he is somewhat quarrelsome, it appears, as kings are apt to be; jealous of any intrusion upon his own territories. A bold proud insect, this Purple High-flier, as he is sometimes called, you see him there in the picture the size of life, and very life-like he looks, with the white patches beautifully relieving the otherwise heavy richness of his dark velvety wings, of which if you were to see the underside, a very different view would be presented; silvery grey, tawny orange, white and black, are there the principal colours, here fading off one into the other, there exhibiting strong contrasts and striking changes of hue. The Caterpillar and Chrysalis of this species are both green, of a fresh vivid tint; the former is marked with pale yellow lines; it swells considerably at the middle of the body, and tapers off at the tail to a point; it is not hairy, and the head is black, with a couple of projec tions like horns sticking straight up, which gives it a very singular appearance. It may be found about the end of May, feeding on the broad-leaved sallow and oak.

The Purple Emperor is chiefly confined to the southern counties of England; we do not hear of its

being taken at all in the north. Yorkshire, Warwickshire, Hampshire, Northamptonshire, Middlesex, Suffolk, Surrey, Sussex, Kent, and Essex, are those
divisions of the country where it has been chiefly
captured; sometimes, as we have already shown, in
great numbers : so that it can scarcely be called a rare
fly. In several of the Kentish woods, not far from
where this little book is written, it may be found at the
proper season, but generally flying high out of the
reach of the net, so that the captures are not numerous ;
the chrysalides are occasionally found, and the Caterpillars sometimes, but very rarely.

BROWN HAIRSTREAK.

PLATE VIII.—FIG. I.

APILIO, or *Thecla Betulæ* is the scientific name of this insect, which is the largest of the five British species contained in the genus *Thecla*—a pretty little family group, by no means remarkable for brilliancy of colour, the upper side of the wings being mostly brown; the under sides, however, are beautifully marked and pencilled with delicate wavy lines, like hair, hence the name Hairstreak, and spots and bands of yellow and white.

One species has the under side nearly all green, on another is a zigzag mark like the letter W; on one the wings approach nearly to a black, and one has a fine purple reflection playing over the brown ground colour. And so we have Purple, Green, White-W, Black, and Brown Hairstreak Butterflies, all of which have a little spike-like lengthening of the lower wings, which distinguishes them from the Argus, and most of the other smaller British flies.

The specific name *Betulæ*, given to the Brown Hairstreak, has reference to the food of the Caterpillar;

Betula, in Latin, signifying a birch tree. This species is not very common in England ; it generally makes its appearance about the end of August or beginning of September, and may be taken mostly near oak or beech trees.

GREASY FRITILLARY.

PLATE VI.—FIG. I.

PAPILIO, or *Melitæa artemis*. The Fritillaries, although by no means the gayest in colour, are among the most beautifully marked of our native Butterflies; there are as many as twelve different species of them, five or six of which only bear the generic name *Melitæa*, derived, it may be, from the Latin *Milites*, a precious stone, of an orange colour— that tint prevailing more or less in all Butterflies belonging to this genus. *Artemis* occurs in the my- thology as the name of a Greek goddess. *Artemesia* is found in ancient history several times; it is also the name of an extensive genus of plants, mostly remark- able for their bitterness, such as wormwood and southernwood, none of which, however, appear to be the common food of the Caterpillars of these elegant flies; that of the Greasy, or, as it is sometimes called, the Marsh Fritillary, feeding on the plantain and the scabious. The first of these names is derived from the peculiar shining appearance of the under side of the fore wings, which look as if they had been oiled; the last from the moist marshy places in which the fly is mostly found. It is generally first on the wing about the middle of May, and may be seen as late as that of July, chiefly in the southern counties of England, but sometimes as far north as Durham and Northumberland.

In Sussex, about Brighton; and Berks, about Enborne, it is said to be particularly plentiful. It measures across the extended wings from an inch and a half to two inches; these wings are of a reddish orange colour, crossed with wavy lines of black, and variegated with patches and spots of delicate straw-colour; the upper and under pair are much alike in their markings; they generally have a dark brown border, and a fringe of silky grey for edging; the under sides are of lighter colours—cream, straw, and silvery grey, with the black lines showing through, and little eyes peeping out here and there, and small crescent-like spangles, and faint waves like clear water or curling smoke, making the prettiest variety that can be imagined.

The Caterpillar is hairy, black above and yellow beneath, with a line of small white dots along the back, and another on each side: it has reddish legs.

The chrysalis is dingy white, with brown spots and

markings; it is generally suspended between several blades of grass, drawn together and fastened at the top. It is not so irregular in shape as many before described, nor is it marked with gold

GLANVILLE FRITILLARY.

PLATE VI.—FIG. II.

APILIO, or *Melitœa Cinxia*. This, although not uncommon on the continent of Europe, is a rare fly in the British Islands. It has been found most plentifully amid the romantic glens and seaward-sloping hills in the Isle of Wight; also in Kent, near Dover and Dartford; and in the counties of Yorkshire, Warwickshire, Lincolnshire, and Wiltshire. May, June, and July are the months in which it may be looked for. It closely resembles the species last described in size and shape, but is somewhat different in colour, and in the arrangement of the markings; the cross lines are more regular, and beautifully waved, and the spots more numerous and distinct; the under sides of the wings are paler, having much of pale straw-colour, the black veins and spots of the upper side being clearly visible.

The Caterpillar of this species is black; it is faintly spotted with white, and has a red head and legs. It feeds on such wild plants as the narrow-leaved plantain, hawkweed, and common speedwell. An English entomologist, named Westwood, says that " these

Caterpillars are found in the autumn living in societies under a kind of tent formed by drawing together the tops of the leaves on which they feed, and covering them with a web."

The chrysalis is pale brown, spotted with orange; in shape much like that of the last species. It is always suspended by the tail end.

The specific name *Cinxia* probably comes from the Latin *Cinctus*—a girdle, having reference to the band-like markings of the wings of the perfect insect, the white dots on the larva, or the row of raised orange spots on the pupa case.

PEARL-BORDERED FRITILLARY.

PLATE VI.—FIG. III.

APILIO, or *Melitœa Euphrosyne.* This is a very beautiful little fly, although it has not much variety of colour: its markings are very regular, or what we should call symmetrical, that is, having a certain accordance or agreement with each other; veins, and spots, and crescents, and dashes of black, upon a deep orange ground, all as regular and even as though they had been traced out by means of rule and compass, and then carefully filled in by a very steady and practised hand, and all round the outer edge runs what seems to be a silken string of tiny pearls. It is indeed an exceedingly pretty fly, with its under wings like embroidered satin of yellow and pale brown, with the faintest flush of red, all so curiously marked, and veined, and mottled. You may see it flitting about towards the end of May, and again quite late in the autumn, when it appears that a second brood issues from their prisons, the shape and colour of which have not been exactly ascertained, to sport awhile in the sunshine, spreading their wings to the extent of perhaps two inches across, although more frequently an inch and a half is the utmost reach of the downy membranes. This fly may be found in most parts of England, as well as Scotland; it is by no means a rare species, although not so common as many.

The Caterpillar feeds on the leaves of different species of violet. It is black and hairy, with a double row of orange spots along the back.

The specific name *Euphrosyne*, carries us again to the ancient mythology, where we find it applied to one of three sisters, so beautiful that they were termed the Graces.

> Fair and graceful they might be,
> Stepping lightly as a fairy;
> But they could not sport like thee
> In the sunshine—Fritillary!

WEAVER'S FRITILLARY.

PLATE VI.—FIG. IV.

APILIO, or *Melitœa dea*. This it will be seen is a smaller fly than the species last described; it is also much rarer, only a few specimens having been taken in this country: the first of these to which the attention of scientific entomologists was called, was captured in Sutton Park, near Tamworth, in Staffordshire, by a Mr. Weaver, by whose name the insect was called. It is a remarkably elegant little fly, resembling closely in its tints and markings *Miss Euphrosyne*, to whom we have just dedicated a line or two.

The Caterpillar, which is dark-coloured and hairy, feeds upon the leaves of the sweet violet; the chrysalis has not been described, probably because not found by a British Naturalist. There are said to be two broods in the year.

The specific name seems to come from the Latin *di* or *dis*, meaning something which differs from another, as this fly does, although but slightly, from the better known preceding species.

1

2

1. Brown Fritillary.   2. Venus Fritillary.

HIGH BROWN FRITILLARY

PLATE VII.—FIG. I.

APILIO, or *Argynnis adippe.* There are but five species of British Butterflies included in the genus *Argynnis;* these are all Fritillaries, and are the largest and most richly ornamented of that family group; the generic name comes, no doubt, from the Latin word *argent*—silver, as these flies are especially remarkable for the large spots and streaks of beautiful silvery white with which the under sides of the wings are decorated.

The High Brown is a fine species of Fritillary, measuring as much as two inches and a half across the expanded wings. It is found in considerable numbers in most of the southern counties of England, and has been taken as far north as Nottinghamshire. It appears about the end of June, or beginning of July; and commonly frequents heaths and the borders of woods.

The upper side of the wings is a rich red brown, with a greenish tinge at the base, that is, where they are united to the body, which is also red brown. Lines, and dots, and crescent-shaped waves of black, arranged with the utmost regularity, give to the whole a tesse-

8

lated or chequered appearance, as though it were a piece of inlaid work. We have in this country a wild plant — a kind of lily — the leaves of whose flesh-coloured blossoms are in like manner chequered with black spots; the old English name of this plant is Fritillary, and in the resemblance, real or fancied, of its blossoms to the wings of the Butterflies so called, originated their family title.

The Caterpillar of this species feeds upon the leaves of the sweet violet, and upon those of its near relative the pansy; no doubt also upon several other plants, although this has not been clearly ascertained. It is of a reddish colour at first, but turns to olive green after the first or second change of skin; it has an interrupted white line along the back, and a row of longish white spots on each side. It is hairy, but not very thickly covered.

The chrysalis is of a reddish colour, with silvery spots; the insect remains in this stage of its existence about a fortnight.

VENUS FRITILLARY.

PLATE VII.—FIG. II.

APILIO, or *Argynnis Aphrodite.* This is at once the rarest and most beautiful of all the Fritillaries with which we are acquainted, and the name of the ancient goddess of love and beauty has therefore been bestowed upon it. *Aphrodite* is but another title of that fabulous personage; it comes from a Greek word signifying froth, because she is said to have been born of the froth of the ocean—a frothy kind of story altogether, and the sooner it is blown away by the wind of truth the better.

Of this Venus among Butterflies, which is properly an American species, but a single specimen appears to have been taken in England, and that was captured in the summer of 1833, in Upton Wood, a few miles from Leamington, Warwickshire. The lucky captor was James Walhouse, Esq.; so says Mr. Morris, in whose large and magnificent volume on ' British Butterflies' will be found depicted the under sides of the wings of this lovely visitant from the "far west;" here is his description of them, as far as words can describe such fair and delicate pieces of Almighty workmanship:—

" Underneath, the ground colour is buff, tinged with pink, the tips greenish, the dark marks showing through. The hind wings are bronze green, but dark at their base, and lighter towards the outside; a row of semicircular silver spots follows the margin, and there are numerous other silver spots."

Of the Caterpillar and Chrysalis of this species, nothing seems to be known. The figure in the plate is the size of life; its elegance of shape, beauty of markings, and rich, harmonious tone of colour all must admire.

1. Brown Hairstreak.   2. Large Copper.   3. Large Blue.

LARGE COPPER.

PLATE VIII.—FIG. II.

APILIO, or *Lycæna dispar*. There are five species of Copper Butterflies, so called from the brilliant coppery hue of the wings, known in Britain, but only two that can be called at all common; indeed, perhaps only one, for the above-named species, since the draining and cultivation of the fenny districts of Cambridge and Huntingdonshire, Norfolk and Suffolk, its favourite places of resort, has in a great measure disappeared.

This fly, the largest of its genus, is remarkable for the flashing appearance of its burnished wings, which measure about an inch and a half across; it flits about among the reeds of the fens and marshes, and sports amidst the rank vegetation of the moist waste lands like a gleam of red fire; when at rest it presents a very different aspect, pale orange and blue ashy grey being the colour of the under sides of the wings; these are diversified with spots of black, encircled with yellow rings and veins, and crescent-shaped marks of the same dark colour. In this fly, as in many other members of the Lepidopterous order, there is a marked difference between the male and female; the latter is considerably the longest, and instead of the wings being entirely of a bright copper colour, except just a narrow black margin and a small, irregularly-shaped spot on each, they are covered with dusky black at the bases, and have wide margins, and many large spots of the same, which greatly detract from the brilliant appearance of the insect.

The Caterpillar is short and thick, swelling up in the middle, and tapering off to a point at each end; the colour is bright green, with a dark line along each side, and little dots like a white powder sprinkled all over it. The food to which it is most partial is a kind of water-dock. The chrysalis, according to Mr. J. F. Stephens, " is at first green, then pale ash-coloured with a black dorsal (that is, back) line, and two abbreviated (shortened) white ones on each side; in shape it bears a close resemblance to a shell twisted and pointed at one end and open at the other."

Curtis says of this splendid fly that " it was first dis-covered in Wales by the celebrated botanist Hudson; it has since been captured in considerable abundance in Wittlesea Mere, Huntingdonshire. July is undoubtedly the right season for this insect, although it is met with in the beginning of August, flying among reeds about the centre of the Mere near Yaxley; it is very active, and in windy weather conceals itself amongst the highest reeds. Upon these the Caterpillar probably feeds, as the Butterfly has been found upon them just emerged from the chrysalis drying its wings."

The origin of the generic name *Lycæna* is somewhat doubtful; *dispar*, in Latin, means unlike, unequal, and may probably refer to the *disparity* of the male and female of this species.

LARGE BLUE.

PLATE VIII.—FIG. III.

APILIO, or *Polyommatus Arion.* The little Blue Butterflies, of which there are as many as nine different species in Britain, must have been observed by most of my readers; they all belong to the genus *Polyommatus*, which word comes from the Greek, and signifies many-eyed; and they are so called on account of the numerous ring-encircled spots resembling eyes, which their wings, especially the under sides, present.

This species is considerably the largest of these pretty flutterers, sometimes measuring as much as an inch and a half or three quarters across the wings, which are of a delicate lilac-shaded blue, having a border of black, as if the fly had gone into mourning for a deceased friend or relative, and just a narrow edging of white, which beautifully relieves the otherwise sombre hue of the whole. The under sides of the wings are a brownish-grey, becoming more blue towards the base. As many as fifty of the black spots, of different shapes and sizes, may sometimes be counted on two wings only.

It is not a common species in England, although it sometimes occurs in particular localities in considerable numbers. Kent, Wilts, Somerset, Bedford, Hants, and Northampton, are the English counties in which it has been captured, as well as in North Wales. The specific name *Arion* is one which occurs in the mythology as that of a celebrated poet and musician, who so delighted the dolphins with his harmony, that when he threw himself into the sea one of them became his horse, and brought him safely to land—very like a whale!

I have now introduced you to one, at least, in several instances to many, of the most remarkable members of the different genera of the Diurnal Lepidoptera, in which British species are included; there is one genus, however, which has no representative in this parliament of Butterflies. This is the Skipper genus, called, in scientific language, *Hesperia*, probably from Hesperus, the evening star; they are all small brown, or orange-coloured flies, variously chequered and marked; there are seven British species, mostly found in and about woods. The genera *Pontia* and *Mancipium* are likewise unrepresented here; to the former belong the white flies, which are so common at certain seasons of the year. The wings of some of these are delicately veined with black, and sometimes with green, and they are all worthy of a close inspection. In the latter genus we find the pretty Orange Tip, the Large Black-Veined or

Hawthorn Butterfly, and several other elegant species, the names of which will be found in the following list, containing those of all the kinds known to be British. It is our purpose, in a future volume, to give a similar list of the Nocturnal and Crepuscular Lepidoptera, that is, the Moths of this country, with illustrations of the most beautiful and remarkable species. Foreign Butterflies and Moths will also form the subjects of two other volumes, so that four of them will contain a tolerably comprehensive account of the great Lepidopterous family, and induce, as we hope, many young readers to pursue the study of entomology in a regular and scientific manner. We wish these little books to be regarded as so many ornamented doorways, leading into the great temple of nature; and we would invite students to pass through and examine the wonders and beauties which lie beyond, praising God the while for the wisdom and goodness displayed in all His works, so various and so wonderful.

The study of insects has sometimes been objected to as trifling and frivolous, but no careful examination of aught which has been created by the hand of the Almighty can be so; and the smaller the creatures examined, the more striking are the proofs which we meet with of the exercise of a power and a skill such as man never possessed.

Let us, then, say with Cowper—

> " How sweet to muse upon His skill displayed,
> (Infinite skill!) in all that He has made ;
> To trace in Nature's most minute design
> The signature and stamp of Power Divine ;
> Contrivance exquisite expressed with ease,
> Where unassisted sight no beauty sees ;
> The shapely limb and lubricated joint,
> Within the small dimensions of a point :
> Muscle and nerve miraculously spun ;
> His mighty work who speaks, and it is done :
> The invisible in things scarce seen revealed ;
> To whom an atom is an ample field."

LIST OF SPECIES.

A COMPLETE LIST OF BRITISH BUTTERFLIES.

With the most common Scientific Name of each, the Time of its Appearance, and usual Places of Resort.

ENGLISH NAME.	SCIENTIFIC NAME.	TIME OF APPEARANCE.	PLACES OF RESORT.
Swallow-tail,	Papilio Machaon,	May to August,	Fens and marshy places.
Scarce Swallow-tail,	Papilio Podalirius,	May to August,	Fens and marshy places.
Brimstone, or Sulphur,	Gonepteryx Rhamni,	February to September,	Gardens, lanes, and fields.
Clouded Yellow,	Colias Edusa,	August to October,	Clover fields and grassy cliffs.
Pale Clouded Yellow,	Colias Hyale,	May to October,	Clover fields and sunny banks.
Black Veined,	Pieris Cratægi,	June to September,	Heaths and forest lands.

A Complete List of British Butterflies—continued.

ENGLISH NAME.	SCIENTIFIC NAME.	TIME OF APPEARANCE.	PLACES OF RESORT.
Large White,	Pontia Brassicæ,	May to September,	Gardens, fields, and lanes.
Small White,	Pontia Rapæ,	April to September,	Gardens, fields, and lanes.
Green Veined,	Pontia Napi,	April to September,	Gardens, fields, and lanes.
Chequered, or Bath White,	Pontia Daplidice,	April to September,	Woods and fields.
Wood White,	Pontia Sinapis,	April to September,	Woods and heaths.
Orange Tipped,	Mancipium Cardamines,	May to August,	Woods and gardens.
Marbled White,	Hipparchia Galathea,	June to September,	Cliffs, woods, and moors.
Wood Argus,	Hipparchia Ægeria,	April to September,	Woods and lanes.
Wood Ringlet,	Hipparchia Hyperanthus,	June to August,	Woods and lanes.
Gate-keeper, or Speckled Wall,	Hipparchia Megæra,	May to August,	Lanes and park walls.
Rock-eyed Underwing, or Grayling,	Hipparchia Semele,	July to September,	Downs and heaths.
Small Meadow Brown,	Hipparchia Tithonus,	July to September,	Lanes and meadows.
Large Meadow Brown,	Hipparchia Janira,	June to August,	Lanes and meadows.
Heath, or Small Ringlet,	Hipparchia Davus,	June to August,	Heaths and commons.

A Complete List of British Butterflies—continued.

ENGLISH NAME.	SCIENTIFIC NAME.	TIME OF APPEARANCE.	PLACES OF RESORT.
Least Meadow Brown, or Small Heath,	Hipparchia Pamphilus,	June to October,	Heaths and meadows.
Arran Argus,	Hipparchia Ligea,	Not exactly known	Isle of Arran.
Scotch Argus,	Hipparchia Blandina,	Not exactly known,	North of England, and South of Scotland.
Small Ringlet,	Hipparchia Cassiope,	June to August,	Mountainous districts.
Silver-bordered Ringlet,	Hipparchia Hero,	Not known—very rare,	Wood sides.
White Admiral,	Limenitis Camilla,	July to September,	Woods and forests.
Red Admiral,	Vanessa Atalanta,	September and October; also in the spring,	Gardens and fields.
Peacock Butterfly,	Vanessa Io,	July to September, and in the spring,	Gardens and fields.
Large Tortoise-shell,	Vanessa Polychloros,	July to September, and in the spring,	Woods and pastures.
Small Tortoise-shell,	Vanessa Urticæ,	July to October, and in the spring,	Gardens and hedge-rows.

A Complete List of British Butterflies—continued.

ENGLISH NAME.	SCIENTIFIC NAME.	TIME OF APPEARANCE.	PLACES OF RESORT.
Camberwell Beauty,	Vanessa Antiopa,	August and September,	Gardens and fields.
Comma Butterfly,	Vanessa C-album,	June to October, and a few in the spring,	Fields and commons.
Albin's Hampstead Eye,	Cynthia Humpstediensis,	Only one taken,	Hampstead.
Painted Lady,	Cynthia Cardui,	June to October,	Gardens and fields.
Scarce Painted Lady.	Cynthia Huntera,	But one taken.	
Purple Emperor,	Apatura Iris,	July to	Woods and forests of oak.
Purple Hairstreak,	Thecla Quercus,	July to September,	In and near woods.
Green Hairstreak,	Thecla Rubi,	May to September,	Waste places about bramble bushes.
White-W Hairstreak,	Thecla W-album,	July to	Woods and wolds.
Black Hairstreak,	Thecla Pruni,	June to	Woods and hill sides.
Brown Hairstreak,	Thecla Betulæ,	August and September,	Woods and hill sides.
Duke of Burgundy Fritillary,	Nemeobius Lucina,	June to	Woods and wolds.
Greasy, or Marsh Fritillary,	Melitæa Artemis,	May to July,	Marshes and moist meadows.

A Complete List of British Butterflies—continued.

ENGLISH NAME.	SCIENTIFIC NAME.	TIME OF APPEARANCE.	PLACES OF RESORT.
Glanville Fritillary,	Melitœa Cinxia,	May to July,	Grassy slopes and glens.
Pearl-bordered Fritillary,	Melitœa Euphrosyne,	May to September,	Chiefly woods.
Pearl-bordered Likeness Fritillary,	Melitœa Athalia,	May to September,	Woods, marshes, and heaths.
Small Pearl-bordered Fritillary,	Melitœa Selene,	May to September,	Woods and heaths.
Weaver's Fritillary,	Melitœa Dia,	But two specimens taken.	
High Brown Fritillary,	Argynnis Adippe,	June to	Woods and heathy places.
Dark Green Fritillary,	Argynnis Aglaia,	July and August,	Woods and downs.
Queen of Spain Fritillary,	Argynnis Lathonia,	July and August—rare.	Woods and downs.
Venus Fritillary,	Argynnis Aphrodite,	But one taken.	
Silver-washed Fritillary,	Argynnis Paphia,	July to	Chiefly woods.
Large Copper,	Lycæna Dispar,	July to September,	Fenny districts.
Small Copper,	Lycæna Phlœas,	April to September,	Moist meadows.
Scarce Copper,	Lycæna Virgaurea,	Very rare,	Marshy places.
Purple-edged Copper,	Lycæna Chryseis,	Very rare,	Forest lands.

A Complete List of British Butterflies—continued.

ENGLISH NAME.	SCIENTIFIC NAME.	TIME OF APPEARANCE.	PLACES OF RESORT.
Dark Underwing Copper,	Lycæna Hippothoe,	Very rare,	Meadows and marshes.
Mazarine Blue,	Polyommatus Acis,	Time not known—scarce,	Chalky districts.
Large Blue,	Polyommatus Arion,	July to	Fields and commons.
Holly, or Azure Blue,	Polyommatus Argiolus,	April to August,	About holly bushes.
Little, or Bedford Blue,	Polyommatus Alsus,	May to July,	Chalkpits and wood sides.
Silver-studded Blue,	Polyommatus Argus,	July to	Woods and heaths.
Common Blue,	Polyommatus Alexis,	May to September,	Grassy meads and pastures.
Clifden, or Dartford Blue,	Polyommatus Adonis,	...	Chalky downs.
Chalk Hill Blue,	Polyommatus Corydon,	July to	Chalky downs.
Brown Argus Blue,	Polyommatus Agestis,	June to September,	Chalky downs near the coast.
Grizzled Skipper,	Hesperia Malvæ,	May to July,	Woods and meadows.
Dingy Skipper,	Hesperia Tages,	May to July,	Woods and meadows.
Large Skipper,	Hesperia Sylvanus,	May to August,	Woody districts.

A Complete List of British Butterflies—continued.

ENGLISH NAME.	SCIENTIFIC NAME.	TIME OF APPEARANCE.	PLACES OF RESORT.
Pearl, or Silver - spotted Skipper,	Hesperia Comma,	July and August,	Commons and parks.
Small Skipper,	Hesperia Linea,	July and August,	In and about woods.
Lulworth Skipper,	Hesperia Actæon,	...	Chalky cliffs.
Spotted Skipper,	Hesperia Paniscus,	May to July,	Chiefly woods.

For most of the particulars in the foregoing table, the editor is indebted to 'Morris's British Butterflies, a recently published volume, in which every species is beautifully figured, and well described by a thoroughly practical entomologist.

INDEX OF SPECIES

FIGURED AND DESCRIBED IN THIS VOLUME.

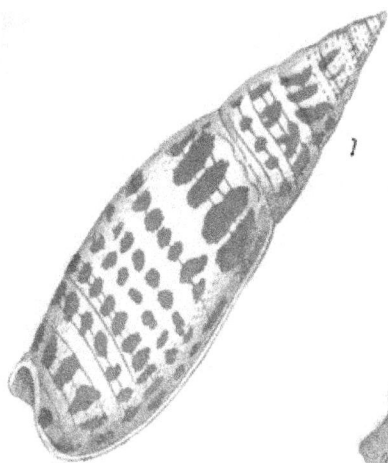

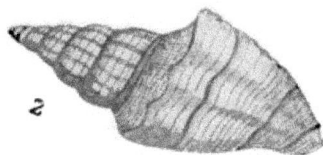

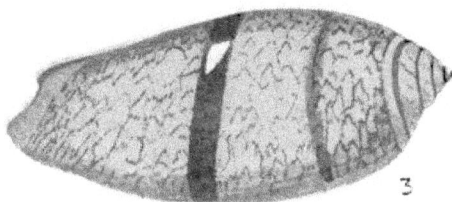

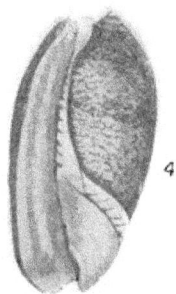

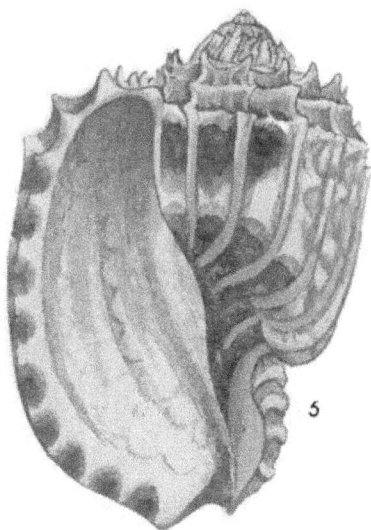

BEAUTIFUL SHELLS

THEIR

Nature, Structure, & Uses Familiarly Explained

WITH

DIRECTIONS FOR COLLECTING, CLEANING, AND ARRANGING
THEM IN THE CABINET,

AND DESCRIPTIONS OF

THE MOST REMARKABLE SPECIES.

BY H. G. ADAMS,

*Author of "Nests and Eggs of Familiar Birds," "Beautiful Butterflies,"
"Humming Birds," "Favourite Song Birds," etc., etc.*

Illustrated.

LONDON:
GROOMBRIDGE AND SONS.

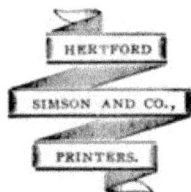

HERTFORD
SIMSON AND CO.,
PRINTERS.

CONTENTS.

UNIVALVES.

BIVALVES.

MULTIVALVES.

BEAUTIFUL SHELLS.

WHAT ARE SHELLS?

DR. JOHNSON gives us no less than eight different meanings for the word SHELL. First, he calls it "The hard covering of anything; the external crust." Second, "The covering of a testaceous or crustaceous animal." And here we may stop, for this is just the signification which has to do with our subject; so let us turn the sentence inside out, and see what we can make of it. We all know what a covering is—an outer coat, a case, a protection from injury, a husk, a crust, a—in short, a shell,—*scyll* or *scell*, as our Saxon forefathers called it; *schale*, as the Germans now term it. No Latin nor Greek here, but the good old Saxon tongue, somewhat rough and rugged, perhaps, but stout and sturdy, and honest and serviceable; a kind of language to stand wear and tear, like a pair of hobnailed shoes, with little polish, but useful, yes, very useful! Well, we have got so far, now comes a hard word—TES-TA-CE-OUS, what can it mean? It is pronounced *tes-ta-shus*, comes from the Latin *testaceus*—having a shell, and means consisting of, or composed of shells; so we find that a *testacean* is a shell-fish, and *testaceology* is the science of shells. Johnson's second meaning of the word *testaceous* is "Having continuous, not jointed shells, opposed to crustaceous." So we find that some naturalists call those testaceous fish, "whose strong and thick shells are entire and of a piece, because those which are joined, as the lobsters, are crustaceous."

Now some of the true testaceans have shells in more than one or two pieces, and therefore this last explanation of the term is rather calculated to mislead a learner; but we shall explain presently wherein the difference consists between them, and the CRUS-TA-CEOUS, or, as we pronounce it, *krus-ta-shus*, fish consists. Here is another long word, it comes from the Latin *crusta*, a word of many meanings, all having reference to an outer coat or covering. My readers know all about pie-crust, and have perhaps heard a surly, snappish, peevish person called a crusty fellow; they will now understand what is meant by a *crustacean* and *crustaceology*, that part of Zoology which treats of crustaceous animals. They constitute quite a large family, these *ologies*, and have a strange way of twisting themselves about, and exchanging limbs and features, so that one is puzzled at times to tell which is which. But here we have fixed two of them, called TESTACEOLOGY and CRUSTACEOLOGY, twin brothers, and very much alike in their characteristics. Let us have a good look at them, so that we may know them again if we should lose sight of them for awhile. Now we will spell over the name of the first—

CONCHOLOGY.

Why, it is changed already! Has this science of shells then another name? Yes, and this is it, pronounced *kong-kol-o-gy*, and derived from the Latin *concha*, which means properly a shell - fish with two shells, joined by a hinge, as the oyster, the cockle, etc. This present volume then is a work on *Conchology*, the subject of it is *Conchiferous*, and whoever studies it will be doing something towards becoming a *Conchologist*.

So much for names and titles; but still we have the question to answer, What are shells? In a learned work called a Cyclopædia, we find it stated that "shell is the hard calcareous (that is chalky) substance which protects, either partially or entirely, the testaceous mollusks externally, or supports certain of them internally." All this you will understand, except perhaps the word " mollusks "; this is a term applied to soft-bodied animals, such as shell-fish, snails, etc., about which we shall have more to say as we proceed. By this we learn that *all* shells are not external or outer coverings, some are internal or inner supports for the soft jelly-like bodies to which they belong, thus performing the duty of bones. An example of this is seen in the shell of the Cuttle-Fish, called by naturalists *Sepia*, a description of which will be found further on in the book.

Shells are either *Crystalline* or *Granular*. Now look at those two words, they almost explain their own meaning. Crystalline shells are those which have more or less of clearness, transparency as we say, so that if held against the light it shows through them; they are sometimes called *Porcellaneous* shells, from their resemblance in this respect to porcelain, or chinaware; the Common

Cowry (*Cyprœa Tigris*) is a shell of this description.

Granular, or, as they are sometimes called, *Concretionary* shells, are the most hard and compact; it is in these that the substance called *nacre*, or *mother-of-pearl*, is mostly found. One of the commonest examples is the oyster shell; if broken across it will be seen to consist of very thin plates, or *laminæ*, as they are termed, closely packed together. The thinner these laminæ may be, the more lustrous and beautiful appears the lining of the shell; that shifting play of colours which we call *iridescent*, from *iris*, the rainbow, is then brightest and most noticeable. A very remarkable substance is this mother-of-pearl; smooth, and shining, and delicately-tinted. Who would expect to find such a beautiful lining to the rugged, rough, dingy-looking oyster, or mussel shell? Truly these mollusks, some of them, live in gorgeous palaces. And the most curious part of the matter is that from the fluids or juices of their own bodies, and from the chalky matter collected from the water, they are enabled to secrete or deposit such wonderfully-constructed habitations, which after all are little more than chalk. Burn a heap of oyster shells, or any other testaceous coverings, and you get lime the same as that produced by burning the white lumps from the chalk-pit, which lumps, by the way, are said to be composed wholly, or for the most part, of marine shells. This we should call *cretaceous* matter, from *creta*, which is the Latin for chalk, or *calcerous*, from *calcis*—lime. Granular shells you have been told are sometimes called *concretionary*, this is because they contain a large amount of this chalky deposit. The rock called limestone, geologists tell us, is composed entirely of fossil shells and mud, or what was once mud, dried and hardened, most likely by extreme heat, to the consistence of rock. Wonderful this to think of; huge mountains, and mighty masses, and far-stretching strata, forming a large portion of the crust of the earth, made up chiefly of the coverings of fishes, a great portion of them so small as to be scarcely visible to the naked eye.—Truly wonderful! But we shall have more to say upon this head when we come to speak of Fossil Shells, as well as on the subject of Pearls, in our chapter on the fish in whose shells they are chiefly found.

It has been a matter of dispute with naturalists whether the testaceous mollusks have shells at all before they issue from the egg, and the main evidence favours the opinion that, generally speaking, they do possess what may be considered as a kind of pattern or model of the habitation which they are to build. This appears to be of a pale horn-colour, and destitute of any markings ; but as soon as the animal enters upon an independent state of existence, it begins to assume its distinctive shape and colour, gradually increasing with the growth of its living tenant, and becoming more and more decidedly marked, until it attains its full perfection of testacean development. Thus the age of some shell-fish can be at once determined by the peculiar conformation and markings of the shell.

The relative portions of animal and earthy, or rather chalky matter, which compose these shells, vary considerably in different kinds ; in those called Crystalline or Porcellaneous, the animal deposit is much less than in the granular or con-cretionary shells, where it not only constitutes a large part of the whole substance, but is more dense, that is, thick, and also has the appearance of being membranous, or organized matter. We can perhaps best explain this by saying that whereas the different chalky layers of the crystalline shell seem merely glued together by the intervening animal fluid ; those of the granular shell, as the oyster, appear to be connected by interlacing mem-

branes. But all this my readers will learn more
about from more advanced and scientific works if
they proceed, as I trust they will do, in the study
of Conchology, a science which has in a greater or
less degree attracted the attention of curious and
contemplative minds in all ages, and the study of
which it has been well said is peculiarly adapted to
recreate the senses, and insensibly to lead us to the
contemplation of the glory of God in creation.

BEAUTY AND VALUE OF SHELLS.

In shells, as in all the works of the Almighty
Creator, we may observe an infinite variety of
form, and if they do not all strike us as alike
graceful, yet in each, however plain and simple,
there is some peculiar beauty, whether it be the
mere hollow cup, or the simple tube, the smooth or
twisted cone, the slender spire, the convoluted oval,
or half circle, ribbed or spiked, with a lip curving
out like the leaf of a water lily, or a narrow rim,
like that of a golden chalice; they are indeed
elegant, each perfect of its kind, and bearing the
impress of a constructive skill far above that of
man, who copies from them some of the most
graceful and elegant designs wherewith to ornament

his buildings, and shapes in which to fashion his articles of luxury or utility.

The most beautiful scroll-work of marble chimney-pieces, cornices of rooms, and other enriched portions of both public and private structures, are those in which the forms of shells have been taken for the patterns of the artistic designs; and how tasteful and appropriate is the employment of the shells themselves as ornaments for the mantel-piece, sideboard, and chiffonnier. Then, too—

> "The rainbow-tinted shell, which lies
> Miles deep at bottom of the sea, hath all
> Colours of skies, and flowers, and gems, and plants."

Not only has it grace and elegance of form, but it has also richness, and delicacy, and variety of colouring. In some species the tints are intensely vivid as the shifting lights of the aurora borealis, or the glowing hues of an autumnal sunset; in others pale and delicate as the first indications of coming morn, or the scarcely perceptible tinge of a just-expanding flower-bud; in some the colours are arranged in patterns, regularly disposed; in others, in masses and blotches, of varying shapes and degrees of intensity; in some again they seem to change and melt one into the other, like the prismatic hues of the rainbow. In all, whether distinct and unconnected, or intimately blended, whether regular or irregular, they are beautiful exceedingly. Nor is their beauty of an evanescent, that is, fading, or vanishing character; unlike plants and animals, which, when once dead, are extremely difficult of preservation, Shells, being composed of particles already in natural combination, are almost indestructible; unless exposed to the action of fire, or some powerful acid, they will remain the same for ages, requiring no care or attention, beyond occasionally removing the dust, which would collect upon and defile their pearly whiteness, or obscure the brilliancy of their colours.

So easily collected, arranged, and preserved, and withal so singular and graceful in form, and rich and various of tint, one cannot wonder that Shells have always had a conspicuous place in all museums, and other collections of natural history objects; neither can we feel surprised that a high value should have been set upon rare specimens; as much as a thousand pounds, it is said, has been given for the first discovered specimen of the *Venus Dione;* another shell, called the *Conus cedo nulli,* is valued at three hundred pounds; and the *Turbo scaloris,* if large and perfect, is worth one hundred guineas; while the *Cypræa aurantium*, or Orange Cowry, if it has not a hole beaten through it, will fetch fifty guineas. It has been calculated that a complete collection of British Conchology is worth its weight in silver.

The following quotation is from "The Young Conchologist," by Miss Roberts. Our readers will do well to peruse it attentively :—"We admit that shells are beautiful, and that they are admirably adapted to the exigencies of the wearers; but how shall we account for the endless diversity of shades and colours, varying from the sober coating of the garden snail to the delicate and glowing tints which are diffused over some of the finer species, in the infinite profusion of undulations, clouds, and spots, bands and reticulated figures, with which these admirable architects enrich the walls of their beautiful receptacles. The means of producing them must be sought for in the animals themselves. Their necks are furnished with pores replete with colouring fluid, which blends insensibly with the calcareous exudation already noticed, and thus occasions that exquisite variety in their testaceous coverings, which art attempts to emulate, but can never fully equal. Thus far is the result of observation and experiment. It now remains to account for the extraordinary fact that the stony exudations of testaceous animals condense only on those parts where they are essential to their welfare. But here investigation ends—the microscope has done its office. It seems as if material nature delighted to baffle the wisdom of her sons, and to say to the proud assertors of the sufficiency of human reason for comprehending the mysteries of creation and of Providence, thus far you can go, and no farther; even in the formation of a shell, or its insignificant inhabitant, your arrogant pretensions are completely humbled.' "

USES OF SHELLS.

In speaking of shells as ornaments, and objects worthy of our study and admiration, we have already mentioned some of their uses, for surely that which contributes to the intellectual improvement and innocent pleasure of mankind, is in its degree useful. But on the more narrow ground of utility, shells may also claim a high place in our estimation. To man in a barbarous and uncivilized state, they furnish the means of performing some of the most important operations of daily life, being extensively used as a substitute for iron. The savage frequently forms his knife, his hunting spear, and his fish-hook of hard shell. Lister relates that the inhabitants of Nicaragua, in South America, fasten a shell, called the *Ostrea virginica*, to a handle of wood, and use it as a spade. In North America the natives use a blue and white belt composed of shells, called the *Venus mercenaria*, as a symbol of peace and unity, and there, too, the gorget of the chieftain's war-dress is composed of the pearl-bearing mussel, called by naturalists *Mytilus margaritiferus*. Many African tribes use the *Murex tritonis* as a military horn, and a rare variety of this shell, which has the volutions reversed, is held sacred, and used only by the priests. Among the Friendly Islanders the Orange Cowry is a symbol of the highest dignity. The Money Cowry (*Cyprœa moneta*) forms the current coin of many nations of Africa ; and a certain number of these shells strung together, are considered by the slave-hunting chiefs as an equivalent for so many black-skinned brothers, whom they sell into hopeless bondage.

Among nations, too, in a high state of civilization, shells are often used for economical, as well as ornamental purposes. To say nothing of mother-of-pearl, which is converted into so many articles, useful as well as pretty, scallop, or oyster shells, are frequently employed as scoops by druggists,

2

grocers, and the like; and in the country the dairy-maid, with the larger kinds of the same shell, skims her milk, and slices her butter; while sometimes by the poor people of both towns and villages, the deeper specimens are converted into oil-lamps. One very important use, my young readers will under-stand, when I speak of a ragged urchin, who shouts to every passer-by—" Please remember the grotto !"

In ancient times, we are told, the people of Athens recorded their votes on public occasions, by marks upon a shell, thus Pope says—

> " He whom ungrateful Athens would expel,
> At all times just, but when he signed the shell;"

in allusion to this custom, of which we are re-minded by such English words as *Attestation*, a certifying, a bearing witness; *Testify*, to give evidence; *Testament*, a will, or written disposal of property, etc.; all having their origin, it appears, in the Latin *testa*—a shell. In ancient poetry, we find the word *Testudo* used to signify a musical instru-ment, also called a lyre or lute; which instrument, according to tradition, was first made by passing strings, and straining them tightly, over the shell of a tortoise. So the poet Dryden, describing

those who listened to the music drawn from this
simple invention, says—

> "Less than a God they thought there could not dwell,
> Within the hollow of that shell
> That spoke so sweetly."

A Greek writer, called Apollodorus, gives this
account of the invention of music by the Egyptian
god Hermes, more commonly known as Mercury.
The Nile having overflowed its banks, and laid
under water the whole country of Egypt, left, when
it returned to its usual boundaries, various dead
animals on the land; among the rest was a tortoise, the flesh of which being dried and wasted by
the sun, nothing remained within the shell except
nerves and cartilages, or thin gristly bones; these
being shrunk and tightened by the heat, became
sonorous, that is sounding. Against this shell
Mercury chanced to strike his foot, and pleased by
the sound caused thereby, examined the shell from
which it came, and so got a notion, as we say, how
he might construct a musical instrument. The first
which he made was in the form of a tortoise, and
strung with the dried sinews of dead animals,
even as are the lutes, harps, and fiddles of our day.
This fanciful mode of accounting for the origin of
music is thus alluded to by a writer named Brown:

> " The lute was first devised
> In imitation of a tortoise' back,
> Whose sinews parched by Apollo's beams,
> Echoed about the concave of the shell;   [sound
> And seeing the shortest and smallest gave shrillest
> They found out frets, whose sweet diversity
> Well touched by the skilful learned fingers,
> Roused so strange a multitude of chords.
> And the opinion many do confirm,
> Because *testudo* signifies a lute."

And now we are among the myths and fables of antiquity, we may just mention another application of the shell to musical purposes. Neptune, who, according to the Grecian mythology, was the god of the sea, is frequently represented as going forth in his car in great state and pomp, with a body-guard of Tritons; some of whom go before with twisted conch shells as trumpets, with which we are to suppose they make delightful harmony. Venus, too, the goddess of beauty, rode on the ocean foam in a testaceous car. Thus Dryden says, that Albion—our native land, so called on account of its chalky cliffs, from the Latin *alba*—white,

> " Was to Neptune recommended;
> Peace and plenty spread the sails;
> Venus in her shell before him,
> From the sands in safety bore him."

But without believing all these fables, more
poetical than true, we may soon convince ourselves
that in the hollow chambers of a shell, there does
seem to dwell, like an imprisoned spirit, a low, sad
kind of music. An English poet, named Walter
Savage Landor, has well described this in these
lines—

> "Of pearly hue
> Within, and they that lustre have imbibed,
> In the sun's palace porch, where, when unyoked,
> His chariot wheel stands midway in the wave;
> Shake one, and it awakens; then apply
> Its polished lips to your attentive ear,
> And it remembers its august abodes,
> And murmurs as the ocean murmurs there."

Wordsworth, too, gives a beautiful description
of a child applying one of these pearly musical-
boxes to his ear.

Many other uses of shells might be mentioned,
to show that they perform an important part in the
operations of nature, as the means and modes by
and in which God sees fit to order the affairs of
this world are frequently called; and also promote
the ends of science, and the arts of every-day life.
By the decomposition of the shells, of which they
are partly composed, solid rocks frequently crumble
to pieces, and spreading over a considerable sur-

face, form a fruitful soil for the nourishment of vegetation. The character of the testaceous deposits, too, enable geologists (as those who study the nature and structure of the earth are termed), to come to important conclusions on many points connected with the subject of this investigation. And if we include, as the subject of our book allows, the inhabitants of shells, how wide a field of usefulness opens before us. How many thousands of our industrious population depend wholly, or in part, upon the capture and sale of shell-fish for their support. In some parts, as the western and northern Islands of Scotland, they have in times of scarcity afforded sustenance to the dwellers on the bleak and barren shores, who but for them must have perished. But of all this we shall have more to say when we come to describe the different members of the testaceous family. We will now offer a few remarks upon

THE INHABITANTS OF SHELLS;

Which belong to that division of Natural History called the *Mollusca*, from the Latin *Mollis*—soft; these *Molluscous* animals, then, are animals having a soft body, and no internal skeleton. You may be quite sure that a *Mollusk* will never break its bones, because it has none to break; it has a shell, however, which may be broken, at least in some cases, for all *Mollusks* have not snug habitations of the kind; but wander about the watery or earthy world in which they live, quite naked; such as the sea and land slugs, and some worms, leeches, etc.; but with these we have nothing to do, our present subject including only a part of

MALACOLOGY,

another member of that queer ology family, deriving its name from two Greek words signifying soft, and a discourse; hence it means a discourse upon soft, or soft-bodied animals, that is *mollusca*.   It is only a part, then, of *Malacology* that we have to do with; that part which relates to the shell-inhabiting mollusks, and strange creatures enough some of these are.   We will have a look at them presently; just now it will be sufficient to observe that the *mollusca testacea*, or soft-bodied animals, furnished with shells, possess the power of exuding, that is, discharging from various parts of their bodies a sticky kind of fluid, which mixing with the chalky matter collected from the water, and becoming hard, forms, in process of time, the shelly covering which is at once a dwelling and a defence for the inhabitant.

Miss Pratt, in her delightful book on " Common Things of the Sea Coast," observes of these shells that, " We gather up those which we find, and looking at their structure would fain know something of the inmate of such a dwelling. All nature proclaims the goodness of God. We hear that the bird which wings its way over our heads has a song of joy; the bee hums delightedly by us; and the little shrimp which darts in the clear pool, seems full of merriment. Was the inmate of the shell less cared for by its Maker? No doubt the little builder had some sense of joy, as he framed from his own substance the house which excites our admiration. Doubtless his existence, short and sluggish as it was, had its own consciousness of pleasure; and obscure as is his history, and little calculated as such a creature might seem to perform an important part in the economy of creation, yet we know that he had a work to do, not only for the living creatures of the sea, but for the well-being of man himself."

CLASSIFICATION OF SHELLS.

The great naturalist, Linnæus, divided shells into thirty-six genera, each of which comprised a number of species; of these species somewhere about two thousand five hundred have been described and classified; the varieties, more or less distinct, are almost countless. Of shells found on and about the British Isles, there are about five hundred and fifty species, or, we should rather say were, for diligent inquirers into this branch of Natural History are almost daily adding to the number.

We have already seen that shells are sometimes called *Crystalline*, and sometimes *Granular*, in accordance with certain peculiarities of construction before mentioned: this is one mode of division. There are several others made use of in different systems of arrangement, which only a deeper study of the subject than can be here entered upon would enable one to understand; the plainest and most common, however, is that which has reference to the form of the shell, which is one of these—

UNIVALVE.   BIVALVE.   MULTIVALVE.

Whelk.   Mussel.   Barnacle.

These words are derived from the Latin, *unus*—one, *bi*—two, *multus*—many, and therefore it may at once be seen that they apply to shells having one, two, or several pieces or divisions. Valve comes from the Latin *valva,* and means a folding door, a lid, a piece moving on a hinge, as the divisions in several of these shells do.

This order of arrangement is generally followed by those who make a collection of shells for a cabinet; and to this we shall adhere, as at once the most simple and convenient, when we come to describe the several species of testaceous mollusks. We will now say a few words on

TAKING AND PRESERVING SHELLS.

A diligent searcher along any beach or coast line will be sure sometimes to light upon curious and valuable specimens, and especially after violent storms may such be sought for, with the greatest chance of success, for the agitation of the waters will then have loosened them from their natural beds and dwelling-places, and cast them on the shore. Very frequently, however, they will be so beaten about and defaced, that they will be comparatively valueless; if enveloped in tangled masses of sea-weed, they are likely to be preserved from injury; and such heaps of uprooted marine vegetation will often afford a rich harvest to the young conchologist, who should always carefully examine them. Many of the shells are so minute as scarcely to be seen with the naked eye, therefore this search can scarcely be properly effected without the assistance of a pocket lens, the cost of which is but trifling. The undersides of pieces of stranded timber, the bottoms of boats lately returned from a fishing voyage, the fisherman's dredge or net, the cable, and the deep-sea line; all these may prove productive, and should be looked to whenever opportunity offers; nor should the search for land and fresh-water shells be neglected, for many of these are very curious, as well as beautiful, and no conchological collection is complete without them. For these, the best hunting-grounds are the ditch side and the river bed, the mossy bank and the hedge-row; amid the twining, serpent-like roots of the old thorn and elder trees; the crevices of the garden wall, the undersides of stones, and all sorts of out-of-the-way holes, nooks, and corners, where may be found the Striped Zebra, and other prettily-marked snail shells, and many other kinds worthy of a place in—

THE CABINET,

which may be either large or small, plain or handsome, in accordance with the means of the collector. Perhaps for a beginner just a nest of shallow drawers in a deal or other case may be quite sufficient; these drawers should be divided into compartments, on the front side of each of which should be pasted a neatly-written label, with the common and scientific name of the species of shell contained in it, together with brief mention of the date when, and the place where, it was taken; and any observations relating to it which can be comprised in a few words, and may seem to the collector of sufficient interest to warrant their insertion. This would be a good beginning; by and by, when the collection is large, the knowledge more ample, and the outlay can be spared, it will be time to think of something ornamental— mahogany and glass, and carved or inlaid work, such a Cabinet as would properly display and enhance the beauties of your testaceous treasures, which are too often stowed away, with other natural curiosities, in a very disorderly, higgledy-piggledy sort of manner, like the collection described by Dryden—

"He furnishes his closet first, and fills
The crowded shelves with rarities of shells;
Adds orient pearls, which from the wave he drew,
And all the sparkling stones of various hue."

When live shells, as they are called—that is, having the living fish in them—are obtained, the best plan is to place them in spirits of wine; this at once deprives the inhabitant of life, without injuring the shell, which should then be placed in hot water for a time; the body of the mollusk is thus rendered firm, and may be removed by means of some pointed instrument. Care should be taken to leave no portion of animal matter within, or after a while it will become putrid, and give out a stain, which will show through and injure the delicate markings of the shell. The surest, most expeditious, and least troublesome mode of cleansing a shell, is to place it in an ant heap for a day or two; the busy little insects will penetrate into its inmost cavities, and remove hence all offending matter. There will be no difficulty in this respect with the multivalve and bivalve kinds, which are only kept closed by means of a set of muscles, which can be tightened or relaxed at the pleasure of the animal within, and become powerless to keep the shell closed as soon as that is dead.

Great care must be taken to preserve unbroken the smaller parts of these shells, such as the hinges or teeth, as on the structure of these the scientific arrangement into genera chiefly depends; the beard, also, as it is called, and silky threads, must not be removed, as these have much to do in determining the particular species.

River and land shells are generally very thin and brittle, and must be carefully handled; their colours are not usually so brilliant as those of the marine species, but they form links in the testaceous chain, which are necessary to a proper study and elucidation of conchology.

The most glowing and gorgeous of all shells are those brought from the Tropical seas, and, excepting in a few rare instances, specimens of most of these can be obtained at little cost from any dealer, or from sailors returning from a voyage. If it is necessary to send either those, or British shells, any distance, or to pack them away in a small compass, the best plan is to wrap them separately in soft paper, place them in a box, and then pour in sawdust, bran, or fine sand, very dry, until all the open spaces are completely filled up.

ON CLEANING AND POLISHING SHELLS.

All shells, whether inhabited or not, when taken, should be soaked for a while in hot water, to remove the dirt which may adhere to them, and dissolve the saline (that is, salt) particles contained in the sea water; they should then be thoroughly dried, and if, as is the case with many, they naturally possess a good polish, they are at once fit for the cabinet. Generally, however, it happens that when shells are dry, they lose much of the peculiar brilliancy of hue which they possess when seen through the medium of the glistening water; to restore this, wash them over with a thin solution of gum arabic, or white of egg. Some collectors use a varnish made of gum mastic, dissolved in spirits of wine; this is, perhaps, preferable, as it is not affected by moisture. Many shells have a very plain, dull appearance, in consequence of being covered over with a kind of skin called an epidermis, a word derived from the Greek, and signifying the outer skin, sometimes called the cuticle. To remove this, soak the shell in warm water for some time, and then rub it over with a stiff brush until the covering is removed; should this be very thick, it will be necessary to mix a little nitric acid with the water; but this must be done very cautiously, for if too strong it will remove all the lustre from the surface of the shell subjected to its influence. Sometimes the file, and a substance called pumice-stone, has to be used, but these are dangerous agents in inexperienced hands. The best polish for the shell, after the skin is removed, is a red earth called tripoli, applied on a piece of soft leather.

FOSSIL SHELLS.

An eminent geologist, named Dr. Mantell, has very beautifully and poetically termed rocks "the Medals of Creation." As on coins and medals we see stamped enduring records of great historical events, so upon the rocks are written by the finger of God a history of some of the mighty changes which the earth has undergone, and fossil shells are among the plainest and most easily read of the characters or letters in which these truths are written. As Dr. Harvey, in his "Sea-side Book," observes, "Shelly-coated mollusca have existed in the waters of the sea, and in rivers, from a very early period of the world's history, and have left in its most stratified rocks and gravels abundance of their shells preserved in a more perfect manner

than the remains of most other animals. Now, as
the species in the early rocks differ from those
found in later formations, quite as much as the
latter from the mollusca of our modern seas, the
gradual change in the character of the embedded
shells marks a certain interval of time in the world's
history." So we see that these rocks are the leaves
of a great book, written all over with wonderful
truths, and those who study such sciences as Geo-
logy and Conchology, are enabled to read much
that is there written.

Every fossil shell that such a student picks out
of the chalk, or limestone, is like a letter in the
Alphabet of Creation; it has a significance, or
meaning, and a number of such put together form,
as it were, words and sentences, that can be made
up into chapters, full of instruction, aye, and of
amusement too. The study may seem a little dry
at first, but never mind, go on, and you will soon
be rewarded for your diligence by the wonders
that will unfold themselves to your understanding
—the fresh and delightful views which you will
obtain into the wide universe, the new and enlarged
ideas of the wisdom and goodness of the Creator,
and of the formation, habits, and connection each
with each of his various creatures.

Properly to treat of fossil shells would require
a book of itself, and a large book too; we can here
but allude to the subject, as a part, and a very
important and interesting part, of the study of
Conchology; more will be said about it in a volume
which we purpose writing for this series, on Rocks,
and the petrified organic remains found in and
about them.

> "Those relics of an older world, which tell
>   Of changes slow or sudden, that have past
> Over the face of Nature; fossil shell,
>   Shark's tooth, and bone of megatheran vast,
>   Turned into stone, and so preserved to show
> Man of those things whereof he ought to know."

UNIVALVES.

GASTEROPODA is a name given by some naturalists to a very extensive group of molluscous animals; the term is derived from two Greek words, signifying stomach and foot;—it has reference to the kind of fleshy foot which generally occupies the whole under side of the body, and by the contraction and extension of the muscles of which, the creature is enabled to glide, with a slow but steady motion, from place to place. The Slug and the Snail are the commonest examples which can be adduced of *gasteropods;* and one may tell by their slimy tracks, shining like silver in the morning sunshine, that during the night, which is their principal feeding-time, they manage to get over a considerable extent of ground, although "a snail's gallop" is a proverbial expression for slow motion; but it is astonishing how much work of any kind may be done if one "keeps at it"; by doing this

the Tortoise beat the Hare, illustrating the truth of
the proverb, that "slow and sure wins the race."
If you watch a Snail travelling with its house upon
its back, it does not seem to make much way, and
you are inclined to think that it will be long ere it
reaches the new settlement to which it appears
journeying with bag and baggage; but leave the
spot for a while, and the chances are that on your
return, the "slow coach" will have got somewhere

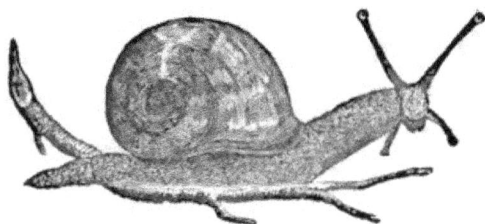

out of sight. Here is a lesson for boys and girls;
whatever you take in hand, don't be in a hurry,
and if people say you are "slow," think of the
Snail, and keep on!

This, then, is a shelled mollusk of the third
class, called *Gasteropoda*, according to the system
of the French naturalist Cuvier. It has a distinct
head, which, like the hinder part of the body,
which we may call a tail if we like, projects, when
the creature is in motion, considerably from the

shell; it is also furnished with what we commonly call horns, naturalists say *tentacles,* from the Latin *tento*—trying, or essaying; with these the creature, as it were, feels its way, being extremely sensitive; they answer the purpose of organs both of sight and touch; put your finger slowly towards one of them, and you will observe that, even before contact, it begins to retract, or draw in, as though sensible of the approach of some opposing body, as it no doubt is. These horns of the Snail, then, are its feelers—eyes to the blind, fingers to the fingerless; so God provides for his creatures all that may be necessary for their existence, and compensates for the deprivation of one sense or organ, by some admirable contrivance which meets the necessities of the case.*

* It appears likely that the little knobs at the end of the Snail's feelers, are, as some naturalists assert, in reality eyes; if so, we were wrong in calling the creature blind. Yet is their position and construction so different from organs of sight generally, that they serve rather to strengthen than invalidate the above observations. The number of the horns varies in different kinds of snails from two to six, and some have none at all. These tentacles, when present, are always situated above the mouth; some of them have the knobs at the base, others at the sides; and it has been conjectured that they may be organs of smell, as well as of sight and touch.

THE COMMON SNAIL

Is called by naturalists *Helix aspersa*, the generic
name being derived from a Greek work signifying
spiral, and having reference to the shape of the
shell; the plural is *Helices*, a term applied to all
convoluted or twisted shells, which terminate in a
point like a church spire: a spiral-shelled fossil is
called a *helicate*. The specific name comes from the
Latin *asper*—rough; whence also our English word
asperity—roughness, and several others. The *Helicidæ*, or *Helix* family, is that which includes the
land shell Snails and the naked Slugs, and in this
family there are several genera; they are distinguished from the shelled water Snails, both sea and
river, by having a different breathing apparatus,
and some other points of internal construction which
it is not necessary to describe here.

The Common Snail has a mouth, of which it
makes good use, as market gardeners well know,
and yet this mouth is not furnished with teeth;
instead of these, the upper lip, which is of a horny
texture, is what is called *dentated*, from the Latin
dentus—a tooth, that is, divided or separated, so as
to present somewhat the appearance of a row of
teeth in the jaw; this lip is of an arched form, and

appears to be a very serviceable kind of instrument to Mr. *Helix aspersa,* who, if his character be not *aspersed,* is very des'ructive to all sorts of greenery. The lower lip is divided only in the middle, where there is an opening of some width : it is not horny, like the upper one.

Snails lay eggs, which are about the size of very small peas; they are soft, and of a whitish colour. Being semi, that is, half, transparent, or clear, their contents can be partly seen ; and in those of a water Snail, deposited against the side of a glass bottle, the young were detected with partially-formed shells upon their backs.

To show how tenacious they are of life, it has been mentioned that Mr. S. Simon, a Dublin merchant, had a collection of fossils and other curiosities left him by his father; among these were some shells of Snails, and *fifteen years* after the collection came into his possession, his son had the shells to play with, and placed them in a basin of water, when lo ! out came the slimy bodies and knobbed horns of several of the *Gasteropods,* no doubt hungry enough after their long sleep.

We all know that our Common Snails hybernate, or sleep through the winter. As soon as the chills of autumn are felt, they seek out some snug crevice in an old wall, or burrow in the earth, or congregate beneath garden pots, roots of trees, thatched roofs, or in any hole or corner that may be convenient, and then throwing a kind of temporary skin, like a drum head, which naturalists call *operculum*, over the opening of their shells, and sticking themselves fast to the sides of their refuge, or to each other, they sleep away, careless of frosts and tempests.

A moist and rather warm state of the atmosphere seems most congenial to the land Snails, some species of which are found in all countries, except those where the most intense cold prevails. Generally speaking, they do not like dry heat, and to escape from it will get under stones, and into other cool places, from whence a shower brings them forth in such numbers, the smaller species especially, as to lead to the popular belief that it sometimes rains Snails.

These *Gasteropods*, although extremely injurious to vegetation, must not be regarded as worse than useless, as they commonly are; besides furnishing food for several wild, as well as domesticated, birds, they are no doubt a nourishing article of diet for man. The Romans had their *cochlearia*, where Snails were regularly fed and fattened for the table;

and the French at the present day their *escargo-toires*, or Snailery, for the same purpose; some of the Snails so kept attain an immense size, as well they may if fed, as by the Romans, on new wine and meal. Many poor persons, especially those who are consumptive, might no doubt derive much cheap sustenance and benefit from using the larger species of *Helicidæ*, which might be collected from hedges and gardens as food. Why should they not eat those as well as the marine mollusks, such as Oysters, Cockles, Whelks, etc.?

Snails have an extraordinary power of re-pro-ducing any part which may be injured or cut off, even to the extent of the whole head, as has been observed to be the case; the reparation of injury done to the shell they can effect easily, as can all testaceous mollusks. Respecting the construction of the shell, it may be observed that it is produced in the thickness of the mantle, or cloak-like covering, which envelops the body of the animal; the forma-tion commences at the small end or spire, and gradually goes on, whorl upon whorl, as the still widening circles which gives the ridgy appearance to most univalves are called. *Columella*, or *pillar*, is the name given to the spire on which the cones are rolled; this is sometimes solid and sometimes hollow; when the latter, the open end is called the *Umbilicus*, meaning the navel or centre. The opening at the bottom, from which the animal issues, is the last portion finished, and this is called the *aperture*, a Latin word adopted into the English dictionary. Some of these *Helices* are wide and flat, even hollow and cup-like, with the whorls rising above the pillar—these are called *discoid* shells;

DISCOID.   TURBINATED.

others which are long and narrow, with projecting spires, are termed *turbinated* shells: the former being more or less flat or disk-like, the latter twisted, whirling, like a spinning-top, from the Latin *turbo*—a whirling, a turning round.

If we take a Common Snail, and plunge it into boiling water, which will instantly kill it, so that it can be removed from the shell, we shall find the whole of that part of the body which was lodged in the upper whorls, or spiral part of the shell, is covered with a thin membrane or skin; this is called the *mantle*, and that portion of it

which corresponds with what we may consider
as the back of the mollusk, and which is considerably thickened, is termed the *collar;* here
are situated the glands, which secrete the colouring
and other matter of which the shell is mainly
composed; although the substance called nacre,
or mother-of-pearl, is secreted in the thinner part
of the mantle; it is however from the collar that
the growth or increase of the shell proceeds. It
is in accordance with certain variations in the
shape and disposition of this mantle and collar,
that shells assume such very different shapes.
Sometimes the whorls or spiral ridges, are pro-
jected or thrown far out, and this produces the
turbinated shell. Sometimes they scarcely rise
above each other, but rather spread towards the
sides, and then we have the *discoid* shape. Generally speaking, the whorls of a shell take a
direction from left to right, but occasionally an
opposite one; they are then called sinistral, or
left-handed shells; such are not common. If one
of the twisted shells be divided lengthways, it
will be seen that the inside of the whorls wind
in an ascending direction, round the *Columella* or
central column, as the spiral staircases in the
Crystal Palace.

But let us return to our Garden Snail, who has many near relatives in Britain, several of which have beautifully-marked and convoluted shells, as will be seen by a reference to our coloured illustrations, Plate I. We will introduce them in due order. Fig. 1, the Banded Snail (*Helix nemoralis*), from the Latin *nemus*—a wood or grove; the prettily-striped shells of this species may be found in great plenty among the roots and in the crevices of the rugged boles of old forest trees, as well as in hedge-rows and in mossy banks, and other situations near woods. Fig. 2, the Heath Snail (*H. ericetorum*), from *erica*, the Latin for heath; a small species with brown bands, remarkable for its large *umbilicus*, perforating the centre of the shell nearly through. Fig. 3, the Silky Snail (*H. sericea*), from the Latin *sericus*—silk-like; the shell of this species is covered with short slimy hairs, which give it a glistening appearance. Fig. 4, the Stone Snail (*H. lapicida*), from the Latin *lapis*—a stone: Linnæus called the species the Stone Cutter, probably on account of its habit of frequenting stony places, and the peculiar construction of the shell, which has a sharp edge running round each whorl; it is commonly found lodged in the cavities of loose-lying stones, but which it can scarcely be suspected of having hollowed out for its own accommodation. Fig. 5, the Elegant Cyclostome (*Cyclostoma elegans*). On turning to the dictionary, we find that *cyclostomous* means having a circular mouth. This species is sometimes called *Turbo elegans*; the beautifully-marked shells are often found in chalky hills covered with brushwood. This pretty mollusk has a curious mode of travelling; the under surface of the foot, which is long, is divided by a deep fissure into two narrow strips, like ribbons; these take hold of whatever the creature may be moving on alternately; one keeping fast hold while the other advances, in like manner to fix itself, and drag the body forward. Fig. 6, the Undulated Plekocheilos (*P. undulatus*); the Latin *plecto*—to twist or twine, seems to be the root from which the generic name of this Snail is derived; the specific name will be easily understood; to undulate, is to flow like waves, and the lines on the shell it will be seen are undulating. This is not a British species, but is introduced here to give variety to the group; it is a West Indian Mollusk, and is found in immense numbers in the forests of St. Vincent; it glues its eggs to the leaves of a plant which holds water, and thus secures for them a damp atmosphere at all times. And here we must conclude our chapter of Land Snails, leaving unnoticed very many beautiful and interesting species, both British and Foreign.

Many poets have alluded to the Snail, but we can only find room for a few verses by Cowper :—

> To grass, or leaf, or fruit, or wall,
> The snail sticks close, nor fears to fall,
> As if he grew there, house and all
> > Together.

> Within that house secure he hides,
> When danger imminent betides
> Of storm, or other harm besides,
> > Of weather.

> Give but his horns the slightest touch,
> His self-collecting power is such,
> He shrinks into his house with much
> > Displeasure.

> Where'er he dwells, he dwells alone,
> Except himself has chattels none,
> Well satisfied to be his own
> > Whole treasure.

> Thus hermit-like, his life he leads,
> Nor partner of his banquet needs,
> And if he meets one only feeds
> > The faster.

Who seeks him must be worse than blind
(He and his house are so combin'd),
If, finding it, he fails to find
Its master.

FRESH-WATER SHELLS.

Many of the following group of Fresh-water Shells, are remarkable for elegance of form, and some for richness of colouring; hence, perhaps, the scientific name applied to the family in which they are mostly included—*Limnœidœ*, which, like *limn*— to paint, agrees with the French *enluminer*. These

mollusks are found in rivers, streams, ditches, and moist marshy places. Like those which live wholly on land, they breathe through lungs, and therefore cannot exist without air; which accounts for their frequently coming to the surface, when under water. In brooks, as well as in stagnant pools, which abound with aquatic plants, they may be found in vast numbers, feeding upon the moist vegetation.

The Common Limnea (*L. stagnalis*) is mostly an inhabitant of stagnant waters, where it is often seen floating with the shell reversed, as in a boat; this shell, like most of those of the Fresh-water Mollusks, is thin, and easily broken; the shape, it

will be seen, is peculiarly elegant, the spire being slender and pointed—very different from that of the Spreading Limnea, called by naturalists, *L. auricularia*, from *aurus*—the ear, to which the broad aperture, or opening of the shell, may be compared; this resembles the other species in its habits. The Horny Planorbis, in Latin *P. corneus*, from *cornu*—a horn. The shape, you will see, is flat, the whorls rolling upon each other like the

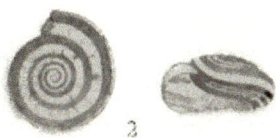

2

3

4

5

Plate I

folds of a bugle horn; this shape would be termed orbicular, from *orbis*—a sphere, or circular body. This is the largest European species of Fresh-water Shells so constructed; it is often found in deep clear ditches, and yields a beautiful purple dye, which, however, soon becomes dull, and changes; it cannot be fixed, and is therefore valueless. The mouth of this shell, in fine specimens, is tinged with pale violet or lilac.

There is another kind, the Keeled Planorbis (*P. lurinatus*), which has the outer edge of the shell

finely ridged, or keeled; it is very small, and very plentiful in fresh-water, both running and stagnant; where, too, is found the Common Physa (*P. fontinalis*), the latter word meaning a spring or fountain. This little mollusk is a quick and active traveller, it sometimes comes out of its shell and throws itself about in an extraordinary way, keeping fast hold by its foot; the generic name, *Physa*, would seem to have reference to the round,

4

smooth, delicate shell, and to come from the same root as *Physalite*, which means a topaz : the members of this genus are very numerous, being found nearly all over the globe. The next belongs to the family *Auriculadæ*, or Ear Shells. The Midas' Ear (*A. Midæ*), this handsome shell is prized by collectors ; it comes from the East Indies. Midas, it is said, was one who set himself up for a judge of music in the old fabulous times, and not appreciating that of Apollo, was rewarded by the angry god with a pair of ass's ears.

The Cone-shaped Melampus (*M. corniformis*), also an Ear Shell, is found in the rivers of the Antilles Islands. It is a pretty shell; the formation is much the same as that of many of the most highly-prized varieties of Marine Shells ; of these we shall have to speak presently. *Melampodium* in Latin, signifies a poisonous plant called Black Hellebore ; in the Mythology, *Melampus* was a great magician, who did all sorts of wonderful things; but we cannot tell what relation there exists between either the plant or the magician and this pretty cone shell. To give variety to this group, we will now throw in a land species called *Megaspira Ruschenbergiana*, about the origin of whose name we cannot even hazard a guess; the termination of the generic name, you will see is *spira*, and a glance at the shell will at once suggest a reason for this; its long tapering spire consists of twenty-three closely-set gradually increasing whorls. This is a rare shell, whose

inhabitant has not yet been described by naturalists; several of the marine species closely resemble it in shape. Much more might be said about the Land and Fresh-water Shells, but we must here leave them, having a wide field before us—namely, the Sea or Marine *Testacea*, one of the most common of which is

THE WHELK,

A univalve shell inhabited by a gasteropod mollusk, or, we should rather say, naturally so tenanted, for very frequently it is taken possession of by the Soldier or Hermit Crab, which having no hard covering to protect their soft plump bodies, are obliged to take lodgings where they can get them, and generally prefer the Whelk shell, of which we here give a figure.

This is one of the commonest of our Marine

Mollusks; it is called by naturalists *Buccinum un-datum;* the first, or generic term, being the Latin for a trumpet, and the second, or specific name, meaning waved, or, as we often say, undulated. So we call this the Waved Whelk; fishermen term

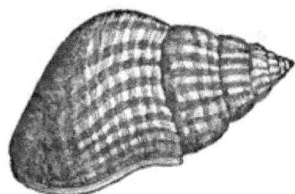

it the Conch, or the Buckie, and tell strange stories of its ravenous appetite and murderous propensities; how, with its spiny tongue, situated at the end of a long flexible proboscis or trunk, it drills a hole in the shell of the Oyster, or other testacean, and sucks out the contents; empty shells, so drilled, are frequently found on the shore, and often, when the dredge is let down into an oyster bed, it comes up time after time filled with Whelks, of which such numbers are sometimes taken, that they are sold to the farmers to be used as manure for the soil. This mollusk is a favourite article of food with the poorer classes of our land, but it is hard and indigestible. The shell may frequently be found in large numbers among the beach stones;

it is strong and firm, from three to four inches long, of a dirty yellowish white. There are two other Whelks common upon our coasts—the Stone or Dog Whelk (*B. lapillus*), from the Latin *lapis*—a stone; and *B. reticulatum,* so called because the shell is *reticulated*, or marked with many lines crossing each other, like net-work; it comes from the Latin *reticulum*—a net; hence also we have *reticule*—a small work-bag, at one time very much carried by ladies.

———

ROCK SHELLS,

Are so called on account of their rough and wrinkled forms; they are nearly allied to the Whelks, to which they bear a close resemblance. Several species are found on our shores, the most common being the Humble Murex (*M. despectus*), from the Latin *despecto*—to despise; this is often used by the fishermen for bait. Some of the foreign Rock Shells are very curious and beautiful; three of them will be found on Plate II., Figs. 1, 2, and 3. The Common Thorny Woodcock (*M. tribulus*), from the French for trouble, whence we have also tribulation, which is sometimes said to be a thorny path. This curious shell is also called Venus' Comb. It

is found in the Indian Ocean, from whence it is
also brought. Fig. 2, the Woodcock's Head (*M.
haustellum*), from the Latin *haustus*—a draught;
the bill of the Woodcock being adapted for sucking.
This term is also applied to insects that live by
suction. The shell, it will be seen, is destitute of
spines, but it is ribbed and beautifully marked. Fig. 3
is worthy of its name—the Royal Murex (*M. regius*),
from *regno*—to reign. It is a splendid species,
of the rich colouring of which art can give but a
faint impression. It is brought from the western
coast of Central and South America, where, as well
as in the islands of the South Pacific, many new
shells of the genus *Murex* have been discovered.

One shell found on our own coast, often mistaken
for a Whelk, is the Pelican's-foot Strombus, called
in scientific language, *Strombus pes-pelicanus*, which
is but a Latinized form of the English name. This
shell varies greatly in shape in different stages of
its growth, and by an inexperienced conchologist,
the young, middle-aged, and old Strombus might
be taken for distinct species. In the *Strombidæ*
family, so called we know not why, the same word
in Latin meaning a kind of shell-fish, are some
species which have produced pearls. One member
of the family which we sometimes see in collections,

is a large and very beautiful shell; this is the Broad-winged Strombus (*S. latissimus*), probably from *latesco*—to wax or grow broad, or large; *issimus* being in the superlative degree, would indicate that

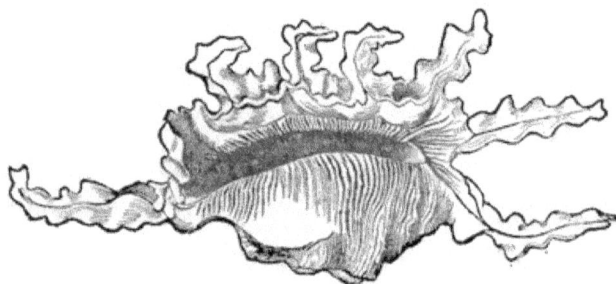

this shell was very much so, as we find it is, some-times measuring as much as twelve inches across. In Plate II. is a representation of this handsome shell, greatly reduced in size, of course. See Fig. 4. We here give a figure, as more curious than beauti-ful, of the Scorpion Pteroceras (*P. scorpius*), which

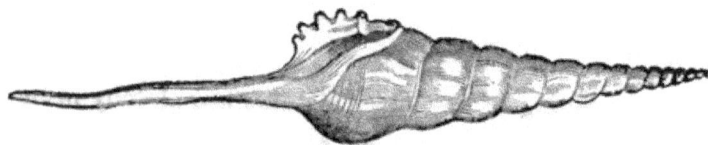

CHINESE SPINDLE.

also belongs to the *Strombidæ* family; as does the curious Chinese Spindle (*Rostellaria rectirostris*). The generic name of the first of these species

comes from the Greek *Ptero*, pronounced *tero*,
meaning a wing, and *cerus*—waxen. Both the
generic and specific names of the second refer to
the peculiar conformation of the shell, being derived
from the Latin, and meaning a straight line or
beak.

On Plate III. will be found the Imbricated Purpura (*P. imbricata*), Fig. 1, which claims a close
alliance with the Whelks. The generic name has
reference to the dye yielded by this, as well as all
the shells of the genus; the specific name comes from
the Latin *imbrex*—the gutter-tile; thus *imbricated*, a
term often used in Natural History, means ridged,
like the roof of a house, where the tiles are placed
to overlap each other, so that the rain will run off.
The Persian Purpura, or, as it is called in Latin,
Purpura Persica, Fig. 2, is another handsome shell
of this family group; its name indicates the place
where it is found. The other species described
comes from South America, and the *P. lapillus*
(the meaning of the specific name has already been
explained), is common on our shores, being found
in great abundance on the rocks at low water.
We read in Scripture of Tyrian purple, and there
is every reason to suppose that the rich colour was
obtained from these and other shell-fish.

PERIWINKLE.

This is the commonest representative which we have of the family *Turbinidæ*, which comprehends, according to Cuvier, all the species which have the shell completely and regularly *turbinated*—that is, if we translate the Latin word into English, twisted. The little Periwinkle (here he is) is by no means a handsome mollusk, but some of his relatives are very beautiful, as we shall presently show. He is called by naturalists *T. littoreus*—from *littoralis*, belonging to the shore —and is often eaten by boys and girls with great relish; but he is not very digestible, and sometimes occasions dangerous disorders. The Swedish peasants believe that when the periwinkle crawls high upon the rocks, a storm is brewing from the south; but Linnæus quotes a Norwegian author to show that according to popular belief, it foretells the approach of a land wind with a calm on shore. Man may learn much of elemental changes from an observation of the movements and habits of all living creatures, which are instructed by God to provide for their safety and wants, and often per-

ceive, long before man himself does, the indications
of calm and tempest, rain and drought, etc. But
our little *Turbo,* what of him? will you boil him,
and pick out his curled-up form with a pin? or let
him go crawling about the rocks, feeding upon the
delicate earlier growth of marine vegetation? In
the former case, you will have to reject the little
kind of horny scale attached to his foot, which
forms, when he retires into his habitation, a closely-
fitting door to make all snug.

Several species of this genus are found on our
shores; one of those is the *Turbo rudis,* or Red
Turbo, which has a very thick periwinkle-like shell,
about three-quarters of an inch long; the colour is
dull red, fawn, or drab.

Of the foreign *Turbinæ,* sometimes called Tur-
ban Shells, we will now introduce two or three
species, which will be found on Plate III. Fig. 3
is the Marbled Turbo (*T. marmoratus*), from the
Latin *marmor*—marble; a large handsome shell
well known to conchologists, and a native of the
Indian seas. Fig. 4 is the Twisted Turbo (*T.
torquatus*); this shell, when deprived of its outer
coat or layer, is beautifully *nacreous,* or, if we may
so speak, mother-of-pearly. The specimens which
have reached England were brought from King

George's Sound. Fig. 5 is called Cook's Turbo (*T. Cookii*); this is a handsome South Sea shell, oftentimes of large size. It has been found in great numbers on the coast of New Zealand.

On Plate IV. we have placed two very curiously formed and marked shells, called Wentletraps, also belonging to the family *Turbinidæ*. The scientific name is *Scalaria*, from the Latin *scala*—a ladder, which the ribbed shells are supposed to resemble. Of this genus there are about eighty distinct species known; they are mostly deep-sea shells found in warm latitudes, although several inhabit the European seas, and one, the Common False Wentletrap (*S. communis*), Fig. 1, may often be picked up on our own shores. Fig. 2, the Royal Staircase Wentletrap, is a rare and valuable shell, generally brought from India and China; the scientific name is *S. pretiosa*, given to it by the French naturalist Lamarck, on account of the high price which it fetched; *pretiose*, in Latin, meaning costly, valuable. As much as £100 have been given for a single specimen of this shell; and a fine one, especially if it exceed two inches in length, yet commands a considerable sum, although not nearly so much as that. A good deal like the False Wentletrap in general outline, is the Awl-shaped Turritella, found in the African and Indian Seas. This is the *T. terebra* of naturalists; the first name referring to the turret shape common to the genus, and the last being the Latin word for an auger, or piercer. The Roseate Turritella (*T. rosea*) is also sometimes seen in collections; the beautiful rosy tint of the live shell changes to a dull red or brown, on the death of the mollusk.

TROCHUS, OR TOP-SHELL.

"Of the shelled Mollusca which the dredge ever and anon brings up," says Mr. Gosse, in his delightful volume on the *Aquarium*, or *Aquavivarium*, as the glass tank in which living marine animals and vegetables are kept, is called, from the Latin *aqua*—water, and *vivo*—to live, "the *Trochi* are among the most conspicuous for beauty. The chief glory of this genus is the richly-painted internal surface of their shells, in which they are not excelled by any even of the true margaritiferous or pearly bivalves."

Of this *Trochidæ* family, a few of the members must be introduced to our readers; it is rather a

numerous one, consisting of more than one hundred
species, which are scattered nearly all over the
world, few seas being without some of them. They
are found at various depths, from near the surface
to forty-five fathoms down, creeping on rocks, sand,
masses of sea-weed, etc. We will first speak of
those found on our own shores, the two commonest,
as well as the smallest of which, are the Grey
and the Spotted Trochi, scientifically named *T.*
cinerarius and *T. maculata*, the translation of the
first Latin specific name being ashy or ash-coloured,
and that of the second, spotted. *Trochus*, in the
same language, signifies a top, and has reference to
the shape of most of these shells, which are something like a boy's whip-top.

Children on the coast sometimes call the last-
named of the above species Pepper-and-salt Shells,
because in colour they resemble the cloth so named.
The Muddy-red Trochus (*T. ziziphinus*), so called,
perhaps, because in colour it resembles the ziziphia,
or fruit of the jujube tree, is also common with us.
This shell is about an inch long, of a grey tint
dashed with dark spots, these follow the line of the
spiral turnings, which are very regular, proceeding
from the opening below to the apex or point. Seen
on shore, its colours are dull and faint, but beneath

the water, inhabited by a living mollusk, it looks as
though made of pearl, and studded with rubies;
the animal, too, is richly coloured, being yellow
with black stripes.   See Plate IV., Fig. 3.

Not so common as the last is another British
mollusk of this genus, called the Granulated
Trochus (*T. granulatus*).   It is the larger, and, as
many think, the more elegant shell of the two,
being in colour a faint flesh tint or yellowish white,
shaded here and there with purple; the spiral lines
which encircle it are composed of small round knobs
which stand out like beads.

There is a singular shell of this genus, called
the Carrier Trochus (*T. phorus*); it is generally
found loaded with foreign objects, such as shells,
small stones, bits of coral, etc., which it attaches
to itself, and so goes about like a collector of
natural curiosities, with his cabinet on his back.

The Imperial Trochus (*T. imperialis*), Fig. 4,
whose scientific name explains itself, is one of the
handsomest shells of the genus; it is very rare, and
has hitherto been found only at New Zealand.   Let
us give our young readers a specimen of the way in
which scientific writers describe shells; thus, this
foreign Trochus, they tell us, is "orbicularly
conical, the apex obtuse, the whorls turgidly

convex, squamose, radiate at the margin." This is quite a simple affair to some descriptions, and simple in fact it is to one, who, by attentive study, has become familiarized with the meaning of the terms. To one also who is acquainted with the Greek and Latin tongues, they will be sufficiently plain, although he has never seen them applied before, for they are all derived from those dead languages, as they are called, and so convey their own meaning to every educated naturalist, no matter of what nation he may be; and hence their chief value. It is not necessary for our readers to trouble themselves about the meaning of such terms at present; by and by it will be necessary for them to do so, if they wish to prosecute the study of any natural science.

But about the Imperial Trochus, with its "orbicularly conical" shell—that term we may explain as round and cone-like; a reference to Fig. 4, Plate IV., will show what is meant by this more clearly than words can, and likewise exhibit the beautiful markings of this species, with its ground -tint of rich violet-brown. This beauty is often obscured by calcareous incrustations and marine plants, showing that the mollusk is sluggish in its habits—a slothful creature. So it is with human

beings, sloth covers and hides the good qualities
and virtues with an overgrowth at all times difficult
to remove, and oftentimes destructive of all that is
fair and good in the character.—Children, be not
slothful! The Obelisk Trochus (*T. obeliscus*), is a
rare white and green shell, sometimes seen in collections; it is of a conico-pyramidal form, not remarkable for beauty, and is a native of the Indian seas.

Mr. Gosse speaks of the Tops and Winkles as
among the most useful inhabitants of the Aquarium;
they mow down with their rasping tongues the
thick growth of *Confervæ*, which would otherwise
spread like a green curtain over the glass walls of
the tank, and obstruct the view of its inhabitants.
Here is this author's description of the beautiful
piece of mechanism by which this work is effected:—
"The appearance and position of the organ would
surprise any one who searched for it for the first time;
and as it is easily found, and as Periwinkles are no
rarities, let me commend it to your examination.
The easiest mode of extracting it, supposing you are
looking for *it* alone, is to slit the thick muzzle
between the two tentacles, when the point of a
needle will catch and draw out what looks like a
slender white thread, two inches or more in length,
one end of which is attached to the throat, and the

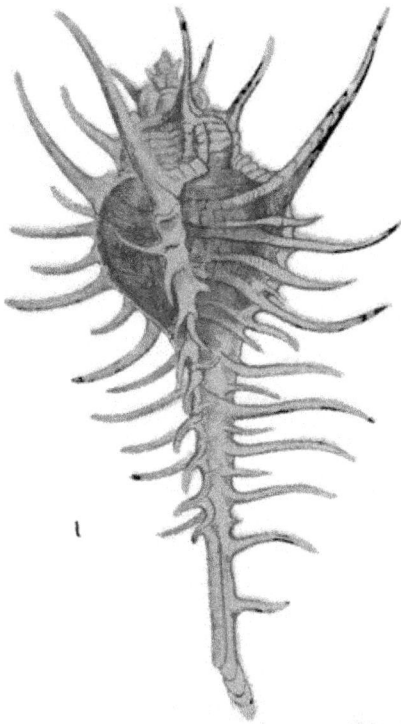

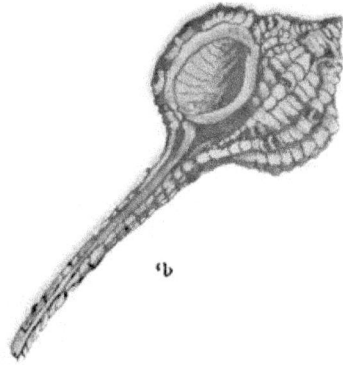

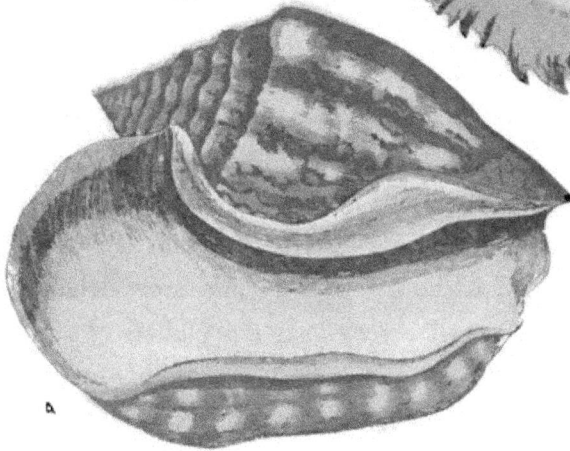

1

'b

3

a

Plate 11

other, which is free, you will see coiled in a beautiful spiral manner, within the cavity of the stomach.

By allowing this tiny thread to stretch itself on a plate of glass, which is easily done by putting a drop of water on it first, which may then be drained off and dried, you will find that it is in reality an excessively delicate ribbon of transparent cartilaginous substance or membrane, on which are set spinous teeth of glassy texture and brilliancy. They are perfectly regular, and arranged in three rows, of which the middle ones are three-pointed, while on each of the outer rows a three-pointed tooth alternates with a larger curved one, somewhat boat-like in form. All the teeth project from the surface of the tongue on hooked curves, and all point in the same direction."

And with this curious piece of mechanism the little Winkle works away and cuts down swathe after swathe of the minute vegetation, just as a mower does the meadow grass; only the mollusk eats as he goes, and so gets payment for his labour; the man has it in another and to him more useful form. We might tell a very long story about these Tops and Winkles, which are nearly related to each other, but must now pass on to describe the rest of

5

the shells on Plate IV., which are the Perspective Solarium (*S. perspectivum*), Fig. 5, the generic name comes from *sol*—the sun, and viewed perspectively, that is, in such a position that the whole top of the shell is at once presented to the view, looking like a flat surface, it presents a circular appearance, marked with rings and rays like representations of the sun sometimes do.

The Variegated Solarium (*S. variegatum*), Fig. 6, is a small but very pretty shell, somewhat rare. The mollusk is remarkable on account of the singular shape of its operculum, which differs from that of all other species; it is of a cone-shape, and covered from top to bottom with what are called membranous lamellæ, that appear to stand out like little shelves winding up spirally. This singular form of operculum has been long known to naturalists, but it is not until lately that they have discovered to what species of testacean it belonged. Let us here explain that *operculus* is the Latin for a cover or lid.

CONES, VOLUTES, MITRES, AND OLIVES.

These are names given by collectors to certain classes of univalve shells, distinguished by peculiarities of formation, more or less distinct. We shall describe two or three of each, that our readers may have some idea of the meaning of the terms which are often used by those who speak or write on conchology.

The family of Cones, called *Conidæ*, is an extensive one; considerably above two hundred species having been discovered. Many of them are very beautiful, both in shape and colour, so that they are highly valued by collectors; they are principally found in the southern and tropical seas, upon sandy bottoms, at depths varying from a few feet to seventeen fathoms. The shells are generally thick and solid, rolled up, as it were, into a conical form; the most familiar illustration that can be given of this form is a sugar-loaf, which all these shells more or less resemble in general outline, as thus—

Cones are either plain or coronated, that is, crowned, having rows of projections round the top of the shell, like the second of the above figures; and this forms a mark of division into two classes, although these classes often run, as it were, one into the other, some plain cones having slight irregularities of surface, and some crowned ones being very nearly plain.

The Common, or Ordinary Cone (*Conus generalis*), Plate V., Fig. 1, is an elegantly-shaped and beautifully-marked shell, having much the appearance of being carved out of some rare kind of marble. The Lettered Cone (*Conus littoralis*), Fig. 2, appears to be scribbled over with Hebrew, Greek, or Arabic characters, and almost every species has something peculiar in its markings; clouds and veins, and dots, and stripes, and bands, of every conceivable shape and mode of arrangement, may be met with in these shells, whose surface, when the epidermis or outer skin is removed, bears a beautiful polish. Curious names have been given to some of them, such, for instance, as the High Admiral, Vice Admiral, and Guinea Admiral, which indicate the rank they hold in the estimation of collectors. From five to twenty guineas is the price at which good and rare ones have been valued, and one, the *Conus cedo nulli,* which may be translated, the Cone second to none, has fetched the enormous sum of three hundred guineas. It must not be supposed that these shells exhibit all their beauties when, inhabited by a carnivorous or flesh-eating mollusk, they move slowly about, or lie for a time motionless among the rocks and sand-beds of the ocean. The before-mentioned epidermis, which is the Latin for the outer skin of the human body, covers them like a cloak or mantle, which is the name it bears among naturalists. Much careful labour is required to bring them to a fit state for cabinet shells.

VOLUTES form an extensive family of shells under the name *Volutinæ.* The greater part are natives of tropical seas, and dwell far down, so that they are seldom found on the coast, except after storms. There are a few European species, but these are not remarkable for beauty, as most of the others are. The generic name signifies twisted, or rather wreathed, as flowers or leaves might be, about some central object. In these shells the spire is generally short, as it is in many cones, sometimes scarcely apparent; the form is usually elegant, and the markings often striking and handsome. On Plate V. will be found three examples—Fig. 3 is the Undulated Volute (*V. undulata*), the Latin for a little wave is *undula*, and these marks are like the lines caused by the flowing of the waves on a sandy shore: this shell is found chiefly in the South Pacific; the animal which inhabits it is prettily marked with zebra-like stripes. Fig. 4 is called the Pacific Volute (*V. Pacificus*), the shape, it will be seen, is somewhat different, being more angular, and it is without the waved lines. Fig. 5, the Bat Volute (*V. vespertilio*), is more decidedly knobbed or spiked, approaching nearly to the shape of some of the coronated ones. This species is found in the Indian seas; the specific name is the Latin for a bat.

MITRES. These are usually considered as a genus, or branch of the Volute family; the scientific name is *mitra*. The form is generally long, slender, and pointed, something like the bishop's mitre, hence the common name of the genus. In the Episcopal Mitre (*M. episcopalis*), Plate VI., Fig. 1, we see this form in its greatest perfection; this is a handsome shell, found in the Indian seas and on the coasts of the South Sea Islands. The mollusk is remarkable for a long proboscis, double the length of the shell, the extremity of which swells into a club form, and has an oval orifice or opening: the

specific name, *episcopalis*, comes from the Latin, and
means of, or like a bishop. The Tanned Mitre (*M.
adusta*), from the Latin *adustus*—burned or parched,
is what is called fusiform and turreted ; that is,
shaped like a spindle, and having a spire or turret-
like termination. The streaks of colour are transverse, that is, running the length of the shell ; or
in other words, they are longitudinal. This, too,
comes from the South Sea Islands. Fig. 2 is the
Wrinkled Mitre (*M. corrugata*), from the Latin
corrugo—to wrinkle. It is very different, both in
shape and markings, from the last species ; the
whorls, it will be seen, are angulated or pointed
above, and the lower part of the shell is much
larger than the spiral or upper portion. It is a
true mitre nevertheless, although not just such a
one as a bishop would like to wear. It inhabits the
Indian Ocean, the coast of New Guinea, etc.

OLIVES. These, for richness of colour and brilliancy of effect, will bear comparison with any
genus of shells. Naturalists speak of them col-
lectively as *Olivinæ*. They belong to the Volute
family, and are said to number about eighty species.
Most of those which have reached this country
have come from the Mauritius, where they catch
them with lines baited with portions of Cuttle-

fish.   We have here depicted two of them, namely,
the Figured Olive, Fig. 3 (*Oliva textilina*), from
the Latin *textilus*, which is woven or plaited ; and
the Ruddy Olive, Fig. 4 (*O. sanguinolenta*), from
sanguis—blood.

We must now bring our notice of the Univalves
to a conclusion.   There are several genera and
many very curious and beautiful species which
we have been unable to notice at all, and of those
which we have, a short account only could be
given—sufficient, however, as we trust, to interest
our readers in the subject, and induce them to
continue the study of it into larger works.   Before
leaving this division of shells, we would call their
attention to one of its greatest ornaments—that
is, the Ventricose Harp Shell (*Harpa ventricosa*),
from the Latin *ventriculus*—the stomach, applied
to this shell on account of its swelled or inflated
shape.   Nothing, however, can be more elegant
than the whole form, nor more beautiful than the
markings of this lovely species (see Plate VI.,
Fig. 5), which belongs properly to the Whelk
family.

COWRIES.

Of Cowries we have already spoken in our chapter on the Uses of Shells. They are among the commonest of our testaceous ornaments, and are remarkable, especially the foreign kinds, for richness and diversity of colour, and the high polish which they bear. The native species are small plain shells, commonly called Pigs, from some real or fancied resemblance which they bear to the swine. They are pretty little white-ribbed shells, and are tolerably plentiful on various parts of the British coasts. There are three kinds—namely, the Louse Pig or Nun Cowry, the Flesh-coloured, and European Pig Cowries. The first of these is of a pale reddish colour, with six square black spots on the back; the second is a beautiful rose tint; and the third is ash-coloured or pinkish, with three black dots and a white streak down the back. The Money Cowry (*Cypræa moneta*), used as current coin in many parts of India, as well as on the coast of Guinea, is a yellow and white shell, with a single band of the former colour; it is small of size, and is sometimes called the Trussed Chicken, for the same reason as the term Pigs is applied to its British relatives. These

Cowries are obtained principally about the Philippine Islands, the Maldive Islands, and the coast of Congo, where, after high tides, the women collect them in baskets, mixed with sand, from which they are afterwards separated and cleaned, when they are ready for the market.

They are only useful as coin so long as they remain unbroken. The value of a single shell is

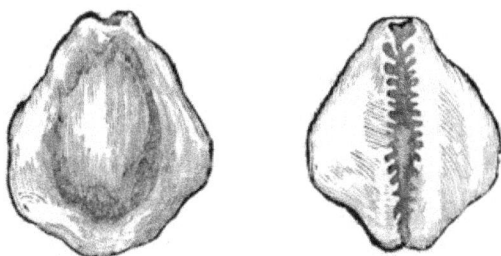

very small, as the following table will show :— Four Cowries make one gunder; twenty Gunders one punn; four Punns one anna; four Annas one cahaun; and four Cahauns one rupee. The value of the latter coin is equal to two shillings and threepence, English money, and this would be exchangeable for five thousand one hundred and twenty Cowries; so that it would never do to pay large sums in this kind of coin: a waggon would be required to convey a few pounds with. In this country the Money Cowries are frequently used as markers or counters in social games; they are generally white, in shape rather broad and flat, being much spread out round the edges, which are slightly puckered like frills. Here are two figures of the shell, exhibiting the back and front view.

On Plate VII. will be found a group of other Foreign Cowries, most of which will be recognized as familiar ornaments of the mantle and sideboard. Fig. 1 is the Spotted or Leopard Cowry, sometimes also called the Tiger Cowry (*O. tigris*), which, in the earlier stages of its growth, is simply marked with broad bands of lighter colour across the shell. Fig. 2, the Map Cowry (*O. mappa*), curiously marked and shaded so as to resemble a coloured map; there are several varieties of this beautiful shell, such as the rosy and dark variety from the Pearl Islands in the Indian Ocean; the Citron and Dwarf Rich-mouthed variety, from the Mauritius. Fig. 3, the Mole Cowry (*C. talpa*), the last word being the Latin for a mole, is of a more slender form than most other species of the *Cypræidæ* family, so called on account of their beauty—*Cyprea* being a name of Venus, the goddess of beauty. Any one who has seen a mole, must be struck with the resemblance of its general outline to this shell, of which there is a darker-coloured variety **of**

somewhat stouter form, called *exustus*—burned or
scorched. Of the Poached-egg Cowries there are
several species, the most common is called by
naturalists *Ovulum ovum,* Fig. 4, from *ovum*—an
egg; the back of this shell is much elevated and
rounded; it is smooth and white; the inside is
orange brown. Some of the Poached-egg group
are of a more slender and angular shape, as, for
instance, that called the Gibbous (*O. Gibbosa*); the
moon when more than half-full, is called gibbous—
that is, rounded unequally, as this shell.

Few shells undergo greater changes,
both of shape and colour, during the
process of growth, than the Cowries,
which are called in France Porcelaines,
on account of their high polish and
brilliant hues; a single species in different stages of development might well be, and
often is, taken for distinct shells. Much might be
said about the Mollusks which inhabit them, but
our present subject has rather to do with their outer
covering than their internal structure. The most
rare and valuable, if not the most beautiful of these
Cowries, is the *C. aurora,* or *aurantium,* Morning-
dawn, or Orange Cowry, a perfect specimen of
which has been sold for fifty guineas. There is a

very curious shell called the Common Weaver's Shuttle (*Oculum volva*), generally included in the *Cyprea* family; of this a representation will be found on Plate VII., Fig. 5. This is brought from China.

BIVALVES.

ACEPHALOUS MOLLUSKS, with Bivalve Shells, is the name given by modern naturalists to the class of animals of which we have now to speak; the only one of these terms which will require explanation is the first; it comes from the Greek, and means headless, so an Acephalan is a molluscous animal without a head, as

THE OYSTER,

Which may be considered as the King of Bivalves; his palace, to be sure, is somewhat rough and rugged outside, but within, its walls are smooth and polished, lustrous and iridescent, and altogether beautiful; of a nacrous or pearly appearance, now flushing into a rose tint, now fading into a pure white, and adorned sometimes with goodly pearls of price; truly this monarch of the *Conchifers* has a habitation worthy of a prince, wherein he lives in right royal state. Our readers may smile, perhaps, at the idea of the solitary Oyster doing this, down there on his mud bank or rocky anchorage ground, shut up in his dirty-looking shells, and holding, as it seems, commune with no one, not even his fellow-mollusks; how can he be said to live in royal state, or, indeed, any state at all, except in a most weary, stale, flat, and unprofitable one? And this only shows how erroneously those often judge who do so hastily, and from first appearances.

If we take a peep through a microscope, under the direction of a naturalist named Rymer Jones, we shall see that "the shell of an Oyster is a world occupied by an innumerable quantity of animals, compared to which the Oyster itself is a colossus. The liquid enclosed between the shell of the Oyster contains a multitude of embryos, covered with transparent scales, which swim with ease; a hundred and twenty of these embryos, placed side by side, would make an inch in breadth. This liquid contains besides, a great variety of animalculæ, five hundred times less in size, which give out a phosphoric light. Yet these are not the only inhabitants of this dwelling—there are also three distinct species of worms."

Let us see if there are any hard names here that want explaining before we go any further. The

first we stumble upon is *Colossus,* which comes from
the Latin, and means a great image or statue, like
that which ancient historians tell us once bestrode
the entrance to the harbour of Rhodes. *Embryo*
comes from the Greek, and means something small
and unfinished, that is to expand or grow into a
more perfect form, as the seed into a plant. *Animalculæ,* are minute or very small animals, such
as cannot be distinguished without the help of a
microscope, hence they are sometimes called microscopic animals; this word comes from the Latin
animalis, which means having life. *Phosphoric*
signifies luminous, or giving out light. The Greek
name of the morning star is *Phospha.* In Latin,
Phosphorus is a term applied to a substance which
chemists extract from bones and other animal
matter, and which, when exposed to air, burns with
a pale blue light, like that emitted by the glow-
worm. Many of the oceanic or sea animalculæ are
exceedingly phosphorescent, so that by night the
waves appear like billows of flame. Of this lumi-
nosity of the ocean, as it is termed, we shall have to
speak on another occasion. We will now return to
the Oyster, who, it will be seen, is by no means so
solitary in his bivalve palace as might be supposed.
He has his torch-bearers, and other attendants,

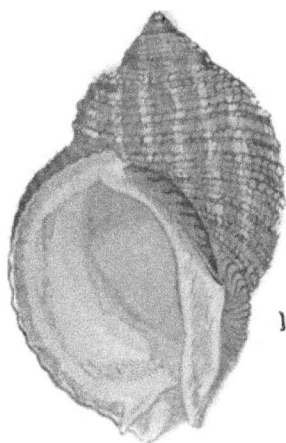

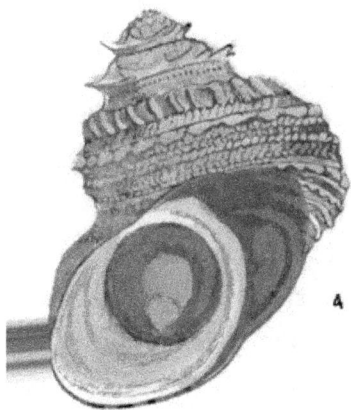

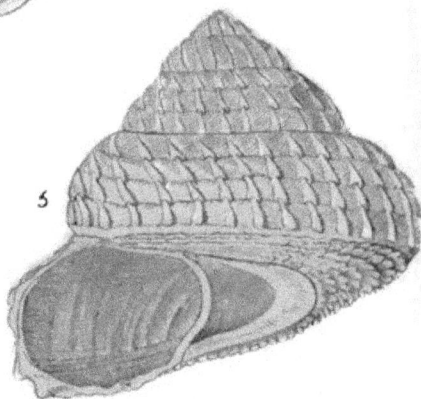

Plate III

quite a host of them, no doubt magnificently dressed, if we could but see them to advantage, and well instructed in the several duties which they have to perform. Oh yes, certainly, as the Irish poet has said,

> " Of all the Conchiferous shell-fish,
> The Oyster *is* surely the King."

Shall we continue the quotation ? and say

> " Arrah Mick, call the people who sell fish,
> And tell them a dozen to bring;
> For it's I that intend to demonstrate,
> The creature's phenomena strange,
> Its functions to set every one straight,
> And exhibit their structure and range."

Scarcely will our limited space permit us to do this, but a few of the most remarkable particulars about this common Acephalan, we feel called upon to set before our readers.

First, then, it belongs to the class *Conchifera.* This is a word which we must stop to examine a little; it seems to come from the Latin *concha,* which means a shell-fish with two shells—in other words, a bivalve mollusk. Second, our Oyster belongs to the class *Pectinidæ.* Now *pecten* is the Latin for a comb, and this class includes those bivalve shells whose edges are toothed, or, as it is said, *pectinated.* In the scallop and the cockle shells this peculiarity is more observable than in those of other members of the class, and these form the typical, or, so to speak, pattern genus, *pecten.* Thirdly, the Common Oyster is a *Monomyarian Conchifer.* Ah! that's something like a name for the acephaloid monarch! Look at these two words, *mono-myarian, di-myarian.* You know, of course, that mony-syllable means one syllable, and dis-syllable means two. You sometimes hear of a person who leads a *monotonous* life, and you think, perhaps, of the Oyster shut up in his shell all alone, *one* by himself. This notion you now know to be a false one, although it is true that he has but one abductor muscle, and therefore belongs to the division of the *Pectinidæ* family called *Monomyaria,* while the Pearl Oyster has two, and therefore belongs to that termed *Dimyaria.* If, as they say, there is reason in the roasting of eggs, surely there must be in the names given to the classes and divisions of shells. We hope to have succeeded in making the why and the wherefore in this case somewhat plain; *one—two—*and away we go out of this maze of hard names. But what about the abductor muscle, above spoken of? Well, this must be explained; *abduce,* coming from the Latin *abduco,* means to separate, to draw away. Hence we have *abduction.* During the life of an Oyster, the usual and natural state of the shell is that of being kept open for a little distance, to allow the water necessary for its nourishment and respiration to flow in and out; but as a security against danger, it was necessary to furnish the animal with the means of rapidly closing the shell, and retaining the valves in a closed state. These actions being only occasional, yet requiring considerable force, are effected by means of a muscular power, for which purpose one or two, or sometimes more, strong muscles are placed between the valves, their fibres passing directly across from the inner surface of one to that of the other, and firmly attached to both; and these are called the abductor muscles, because their office is to draw or pull. How strongly they do this, those whose business it is to open Oysters can best tell. If the animal within were not alive, the process would not be a difficult one, as in that case the muscles would be relaxed, and the shell would come open of itself, so that actually people who eat Oysters directly they are opened, swallow them "all alive-O!"

If a pair of the shells from which the delicious

morsel has been extracted be taken in the hand, it
may be noticed that one is much thinner, smoother,
and flatter than the other: this is the side most
exposed to the action of the water; the rougher
and rounder side is that which is attached to the
rock or other substance to which the animal forms
an attachment that is usually life-long. The two
portions of the shell are joined together by a hinge
of curious workmanship, which is formed of the
inner layer of the shell, and strengthened by a
ligament which is wonderfully elastic; when the
shell is drawn together by the abductor muscles, the
ligament is at full stretch, and as soon as they relax
at all, it contracts, and causes the shell to gape.
This process is repeated as often as may be necessary for the safety and sustenance of the animal.
within, whose mouth is situated at the narrowest
part of its habitation—namely, near the joint of the
hinge which connects its upper and under shell. The
anatomical structure of the Oyster is more perfect
than would be supposed, from its apparently low
state of organization; it has a heart, liver, and in-
testinal canal, and a bag near the mouth, which
answers the purpose of a stomach. Its breathing
organs are gills, closely resembling those of most
other fish; it has little vessels which convey the

bile from the stomach to the liver, and may, perhaps, be subject to bilious attacks as well as those who swallow this curious piece of organization at a mouthful, without thinking at all of the goodly structure they are demolishing. There is the tiny heart with its series of blood-vessels, just as perfect as in the larger animals. There are the nerves in the shape of minute feelers, which appear to be acutely sensible not only of actual contact with foreign bodies, but also of sounds and movements from without. A very nice sense of feeling appears to reside in what is called the beard, in scientific language *bissus ;* this is a kind of double fringe to the two lobes of the mantle, or sac, as it is called, which envelops the body of the animal, and floats free from the shell, except just at the part nearer the valve where it is attached.

We have just spoken of the beard of the Oyster, and this reminds us of a conundrum which may serve to amuse our readers, and enliven these dry details a little. Why is an Oyster the most anomalous—that is strange, contradictory—creature in existence? Do you give it up? Well, then, it is because

> "It wears a beard without any chin,
> And leaves its bed to be tucked in."

Again, by this allusion to the "tucking in" of Oysters, a phrase more expressive than polite, we have recalled to memory the saying of a quaint old uthor, that they are "ungodly, uncharitable, and unprofitable meat: ungodly, because they are eaten without grace; uncharitable, because they leave nothing but shells; and unprofitable, because they must swim in wine." Not, generally, however, are they eaten in this luxurious manner; a little pepper and vinegar is all that they commonly get in the way of sauce, and those who swallow them thus accompanied, seem to do so with infinite relish. A very long chapter, if not a whole book, might be written about the historical associations of Oysters, for which our country has been famous, as far back as the time of the first Roman invasion; much, too, might be said about the Oyster beds and fisheries, which give employment to thousands of our industrious population; but all this has so little to do with natural history, that we can find no excuse for dwelling upon it here. It is quite within the range of our subject, however, to state that the "spat" or "spawn" of the Oyster is cast about the beginning of May: at first it resembles a drop of greenish tallow, but by the aid of the microscope it may be seen to consist of a great number of minute particles, each of which is an egg, and will by and by become a perfect fish; these increase in size very rapidly, and after floating about for a while, sink to the bottom, and become attached to rock or some other substance, in which position, if not violently detached or removed, they complete their growth and live out the term of their natural life. Their food minute animalculæ, and microscopic vegetation, on the nature of which their flavour greatly depends.

They have many enemies besides man; the whelk, and the crab, the sea-star or "five fingers," and the large drum-fish, which swallows them almost by the bushel, shells and all; these help to thin the Oyster-beds, and make the dredger's

labours less remunerative than they would other-wise be. Here is a picture of one as he stands in

his boat just about to throw his dredge into the
sandy bottom, where he knows the delicious testaceans do, or ought to, lie most thickly. The dredge,
which is a triangular iron frame with a net over the
bottom, will naturally sink, and when the line to
which it is attached ceases to run out, the dredger
will put his boat in motion, and draw it thus over
the Oyster-bed, and then pull it up filled, it may be,
with little fat " Miltons," or large " Colchesters,"
or such other kind as the spot is known to yield.

The Latin for Oyster is *Ostrea*, and that is a
name given to a genus of the *Pectinidæ* family,
comprising beside the *O. edulis*, or common Oyster,
many other species. *Edulis* means eatable. Some
naturalists divide these Ostraceans into two groups,
first with simple or undulated, but not plaited
valves; second, those which have the borders of
their valves distinctly plaited.

To the first group belong the Common Oyster,
and between thirty and forty other living species,
which are found principally in warm and temperate
latitudes. In the Polar ocean none have been
discovered, and in the hotter climates they are
most abundant, being found in large beds or banks
near the coast, and often attached to rocks, and
even to trees which grow by the water, so that the

accounts of some old travellers who stated that they saw Oysters growing upon trees, were not so false as many supposed them.

The annexed figure is that of the Cock's-comb

Oyster, *Ostrea Crista-Galli*, a native of the Indian Seas, and a very remarkable shell, on account of its crooked or deeply indented form; the specific name

means cock's-crest. The Chinese Window Oyster, called *Placuna Placenta*, which we may, if we like, translate into a pleasant or agreeable cake, the shell, it will be seen, is round like a cake, and its smoothness and regularity of form render it agreeable to look upon. This species, too, comes from the Indian Seas, where it is taken on sandy bottoms. The American Spiny Oyster, or *Spondylus Americanus*, brings us into another family, that of the Water Clams, called by naturalists *Spondylidæ*;

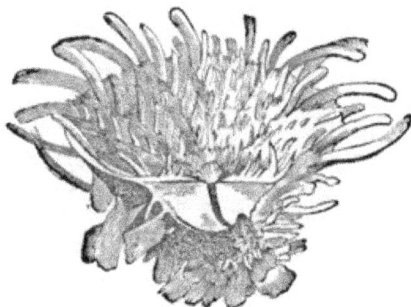

with the spines stuck out every way, and no way in particular, it looks like a head of hair greatly in need of the assistance of one of its *pectinated* relatives. The specific name of this curious shell explains itself; the generic name comes from the Latin *Spondylis*—a kind of serpent.

Passing over the family *Malleidæ*, or Hammer Oysters, we come to the *Meleagrinidæ*, or Pearl Oysters, of which Fig. 1, Plate VIII. is an example.

This is the *M. Margaritifera* of naturalists, the mollusk in whose shells pearls are chiefly found. Here are two long words; *Meleagris* is the Latin for a Guinea or Turkey Hen, to the markings of whose plumage naturalists might have imagined the shells of this genus bore some resemblance. There was, says the mythology, a celebrated hero of antiquity named Meleaga, but we can hardly suppose that there is any association between his name and that of a genus of Oysters, of which edible we read the ancients were very fond, and they are said to have had a fancy, not only for the mollusk itself, but also for the pearls found in its shell, which at their luxurious banquets they dissolved in wine, to make the draughts richer, or, at all events, more expensive; and this brings us to the specific name of the Pearl Oyster, *Margaritifera,* which comes from the Latin *Margarita*—a pearl; the French use this word slightly altered in the spelling (thus, *Marguerite*) for both a daisy and

A PEARL.

This jewel, so highly valued for its chaste beauty, is but a secretion of animal matter, resulting from the efforts of some uneasy mollusk, annoyed by a foreign substance, which has found its way into his habitation, to make the best of an unavoidable evil by enclosing it in a soft smooth covering. Let us imitate the Oyster; and when annoyed or afflicted, by meekness and patience, and Christian charity, strive to turn our vexations and troubles into "pearls of great price," and "goodly pearls," like those mentioned in Scripture.

It is on the north-west coast of the Island of Ceylon, in the Indian Ocean, that the Pearl Oyster most abounds, and there it is that the Pearl fishery is conducted in the most extensive, systematic, and successful manner; this fishing commences at the beginning of March, and upwards of two hundred boats are usually employed in it; in each of these boats are ten divers, who go down to the Oyster-beds, five at a time, and so relieve each other; there are besides thirteen other men who manage the boat and attend to the divers. Altogether, it is computed that from fifty to sixty thousand persons, in some way engaged in the fishery, or preparation, or sale of the pearls, assemble at and near the scene of operations, which must be indeed a busy one. The number of Oysters taken during the period of the fishing, which is about a month, must be prodigious. One boat has been known to bring on shore, in the day, as many as thirty-three

thousand; they are placed in heaps, and allowed
to remain until they become putrid, when they
undergo a very elaborate process of washing and
separating from the shells, which are carefully
examined and deprived of their pearly treasures.
The stench arising from the decomposed animal
matter is described as horrible, and the whole process filthy and loathsome in the extreme; yet out
of the slime and mud and disgusting effluvia, come
every year gems of inestimable value, calculated to
adorn the brow of beauty and form ornaments the
most pure and delicate that can be imagined. For
the exclusive right of fishing on the banks of
Ceylon for a single season, as much as £120,000
have been paid to the English Government by one
person, who sublets boats to others. Pearls vary
greatly in value according to their colour and
size; those which are perfectly white are the most
valuable; next to these are those which have a
yellowish tinge; the smallest kind, used for various
ornamental purposes, are called seed pearls, the
refuse is made into a kind of confection called
chimum, highly relished by Chinese epicures. A
single oyster will sometimes contain several pearls,
which are generally embedded in the body of the
animal, but are sometimes fixed to the shell; it is

recorded of one rich mollusk, that there were found
in his possession no less than one hundred and fifty
precious jewels; he must have been a miser, or
perhaps he had taken them in pledge from his less
provident neighbours.

From the earliest time, pearls have been considered as valuable ornaments; they are mentioned
in the book of Job (see chap. xxviii. verse 18),
and are often alluded to by Greek and Roman
writers.   Various attempts have been made to
imitate them, and one mode of producing them,
practised, it is said, more than a thousand years
ago, is still carried on in China.   In the shells of
Pearl Oysters, holes are bored, into which pieces of
iron are introduced; these wounding and irritating
the animal, cause it to deposit coat upon coat of
pearly matter over the wounded part, and so the
pearl is formed.   Artificial pearls are made of
hollow glass globules or little globes, covered on
the inside with a liquid called pearl-essence, and
filled up with white wax.   Historians speak of an
ancient traffic in native pearls carried on by this
country; and in modern times, British pearls of
considerable value have been discovered—one not
many years since, by a gentleman who was eating
oysters at Winchester, was valued at two hundred

guineas. Generally the pearls of this country are inferior in the two requisites of colour and size.

Interesting accounts of Pearls and Pearl-fishing, will be found in "The Penny" and "Saturday Magazines," and many other works easy of access. There our young readers may learn of the perils and dangers to which the poor divers are exposed from the voracious sharks, which hover about the fishing grounds, and make a dash at their victim, heedless of the written charms with which the priest or shark-charmer has provided him previous to his descent, and of much more than we can find space here to tell. All we can now do is to give the portrait, as drawn by Thomas Hood, of a lady who takes up her abode in all the pearl-producing bivalves, and who is, therefore, perhaps on this account, called

MOTHER OF PEARL.

THE MUSSEL AND THE COCKLE.

It is in the *Dimyaria* division of the *Conchifera* that we must look for those familiar bivalves, the Mussel, or, as it is sometimes spelled, Muscle, and the Cockle; the former, called in scientific language *Mytilus,* which in Latin means simply a shell-fish; and the latter *Cardium,* which may have reference to the hinge of this bivalve, or the heart-shape assumed by several of the species; *cardo,* in Latin, signifying the hinge of a gate, and *cardesco,* a stone in the shape of a heart.

It is to the *Mytilidæ* family that we shall first direct our attention; and here we find the Common or Edible Mussel (*M. edule*), and many other species, in all of which the shell is more or less elongated or lengthened out, and pointed at one end. The members of this family are abundant on most rocky coasts, where facilities are afforded for the mollusks to moor themselves to rocks, stones, and other substances covered at high-water, but left dry by the retreating tide. They are not, however, confined to shores of this description, but are sometimes found in vast numbers on low sandy or pebbly flats, which run far out into the sea; these are called beds of Mussels, and are, like the Oyster

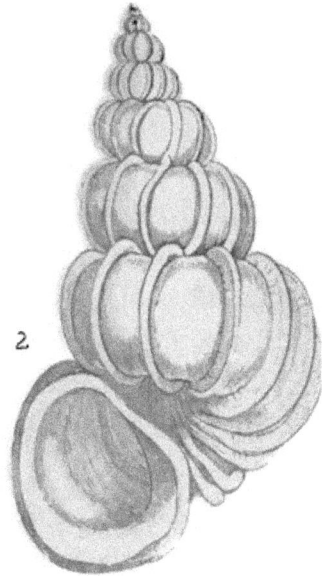

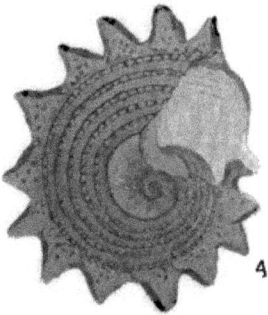

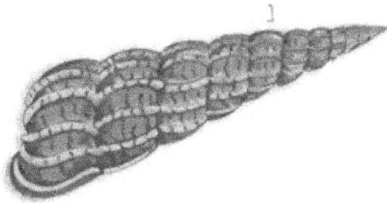

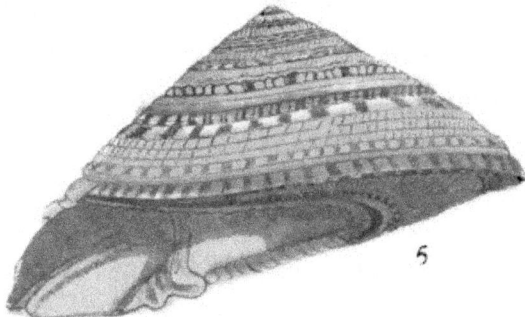

grounds, specially cared for and protected. As a ship by its cable, so commonly the Mussel, by its bissus or beard, is made fast to its anchorage-ground, be it pebbly or sandy beach, or jutting rock. Sometimes, however, the mollusk travels; and this is how it manages to do so : it has a stout, fleshy foot, in shape something like that of a chubby child, and this it can advance about two inches beyond the edge of the shell, then fixing the point of it to a piece of rock or any other body, and contracting it, the shell is drawn onward, and sure, though slow, progress is made in any desired direction. The *Pinna,* as the marine Mussel is called, has a foot which is cylindrical in shape, and has at the bottom a round tendon, almost as long as itself, the use of which appears to be to gather in and retain the numerous threads with which, when inhabiting the shores of tempestuous seas, it lashes itself fast to the fixed objects around; these threads are fastened at various points, and then drawn tight by the animal, whose instinct teaches it that its brittle shell would soon be broken in pieces, if suffered to roll hither and thither at the mercy of the waves.

The Mussel has a very curious method of pre-paring its cable for this service; it is not woven,

7

nor spun, nor drawn out of the body, like the web
of the spider, but produced in a liquid form, and
cast in a mould which is formed by a groove in
the foot, extending from the root of the tendon to
the upper extremity; the sides of this groove are
formed so as to fold over it and form a canal, into
which the glutinous or sticky secretion is poured;
there it remains until it has dried into a solid
thread, when the end of it is carried out by the
foot, and applied to the object to which it is to be
attached; the canal is then opened through its
whole length to free the thread, and closing again
is ready for another casting; as if conscious how
much depends upon the security of his lines, the
animal tries every one after he has fixed it by
swinging itself round so as to put the threads
fully on the stretch. When once they are firmly
fixed, it seems to have no power of disengaging
itself from them; the liquid matter out of which
they are formed is so very glutinous, or glue-like,
as to attach itself firmly to the smoothest bodies.
The process of producing it is a slow one, as it
does not appear that the *Pinna* can form more than
four or five in the course of twenty-four hours.
When the animal is disturbed in its operations, it
sometimes forms these threads too hastily; they

are then more slender than those produced at leisure, and, of a consequence, weaker. On some parts of the Mediterranean coast, as in Sicily, gloves and other articles have been manufactured from the threads of this mollusk. They resemble very fine silk in appearance.

The foot of the Cockle, of which we here give a figure, is commonly employed in scooping out the mud or sand, beneath which it conceals itself; this useful limb assumes the form of a shovel, hook, or any other instrument necessary for the purpose; it appears to be a mass of muscular fibres, and to possess great power. As a boatman in shallow water sends his vessel along by pushing against the bottom with his boat-hook, precisely so does *Mr. Cardium* travel; he doubles up his foot into a club, and by an energetic use of it as a propeller, makes considerable headway along the surface of the soft sand beneath the waters. In this way, too, some members of the genus *solen* force their way through the sand ; while those called *Tellina* spring to a considerable distance, by first folding the foot into a small compass, and then suddenly expanding it, closing the shell at the same time with

a loud snap ; so that you see these sober-looking mollusks are sometimes frolicsome fellows : this is an enforcement of the lesson, judge not by appearances.

Some of the species, both of the Mussel and Cockle families, have very beautiful shells. We give a representation of one of each, on Plate VIII. Fig. 2 is the Magellanic Mytilus, (*M. Magellanicus*,) found chiefly in the Straits of Magellan ; it is generally four or five inches long; the shells when polished are very brilliant, the deep purple colour changing into rich violet, as they are held in different lights. In most cabinets the large fan-like delicate shells of the genus *Pinna* may be observed; the largest species is that called *Pinna flabellum*, taken in the Mediterranean; it sometimes exceeds two feet in length. The first of these names is a Latin word signifying, besides a shell-fish, the fin of a fish, or the wing-feathers of a bird —hence the term pinion; it refers to the fin-like or wing-like shape of this shell. *Flabellum* means a fan, referring probably to the bissus of the mollusk, which is fine and glossy, like silk, and very abundant.

Many pretty specimens for figuring might be selected from the *Naidæ*, a family of Fresh-water

Mussels, so called from the Naiades, fabulous divinities of the streams and rivers. The shells of many of these, which are of considerable thickness, are lined with the most brilliant nacre, and in these, as might be expected, pearls are sometimes found. One species, abundant in some English rivers, called the *Mya Margaritifera*, or, as some say, *Uno elongates*, has long been celebrated for this valuable production. It was most likely with pearls from this mollusk that Julius Cæsar adorned a breastplate, which he dedicated to Venus, and hung up in her temple. The rivers Esk and Conway were formerly celebrated as British pearl fishing-grounds; a Conway pearl was presented by her chamberlain, Sir Richard Wynn, of Gwyder, to Catherine, Queen of Charles the Second; and in the royal crown of Britain this jewel is said still to occupy a place. Sir John Hawkins, the circumnavigator of the globe, held a patent for the pearl-fishery of the River Irt, in Cumberland. The rivers of Tyrone and Donegal, in Ireland, have, or had, their pearl-bearing Mussels; we read of one which weighed thirty-six carats (a carat is nearly four grains), but not being of perfect shape and colour, it was only valued at forty pounds. We also read of another purchased by Lady Glenlealy, for ten pounds, and found to be so perfect and admirable, that eighty pounds was afterwards offered for it, and refused.

These *Naidæ* have not a bissus like the Marine Mussels, they are therefore never attached to one object; they use their foot as a propeller in traversing the muddy floor of the pond or river, and they have a very funny way of getting along indeed; first, they open the valves of the shell, put out the foot, and, after some little hard work, manage to set themselves up on edge; they then proceed by a series of jerks, leaving a deepish furrow in the mud behind them.

We will now go to Fig. 3, the Spined Cytherea, the *Cytherea*, or *Venus Dione* of naturalists; the meaning of the term is the mother of Venus, who was, as you will remember, the goddess of beauty, given to this shell, perhaps, because it is entitled to occupy a place at the head of the *Cytherea*, a genus of the *Cardiidæ*, or Cockle family, of which genus there are about seventy-eight living species; this, as it is the most rare, is also, perhaps, the most beautiful; it is found in the seas of America, and is remarkable for the row of spines on the hinder border of each valve; these vary much in size and number, being in some individuals long and far apart, in others, short, thick, and closely set. The

colour of the shell also varies considerably, being
sometimes of a delicate rose colour ; at others, more
of a claret ;˙ at others again bordering on purple.
It was for one of the first discovered specimens of
this shell that £1000 is said to have been given
Truly a Venus of value this ; it ought to be called
the Queen of Cockles !

Our next examples (see Fig. 4), is the Spotted
Tridacna (*T. maculatus*), the latter term signifying
spotted.   In the *Chamidæ*, or Clam family, is placed
the *Tridacna* genus, the discovered species of which
are not numerous ; they are chiefly found in the
Indian seas.   The one above-mentioned claims pre-
eminence for beauty.   We cannot quite see the
applicability of the generic name; *Tridacnus*, in
Latin, signifies to be eaten at three bites, but he
must be a man of large capacity indeed who could
so devour the head of this family, the Giant Tridacna (*T. gigas*), a single specimen of which has
been known to weigh as much as five hundred and
seventy pounds; from three to four hundred is by
no means an uncommon size.   The shell of this
giant mollusk is of a very picturesque shape, something like its spotted *congener* (as we call anything
of the same kind or genus), only it is somewhat
plainer, and more deeply ribbed and indented.   The

inside is of a glossy whiteness, and it is frequently used as a basin for garden fountains, or the reception of rills or little jets of water, which sparkle in its stainless hollow. In the church of St. Salpice, at Paris, is a shell of this immense Clam, the valves of which are used as receptacles for holy water; it was presented to Francis the First by the republic of Venice. Fancy the clapping to of such a pair of valves when the animal closes its shell in alarm, and

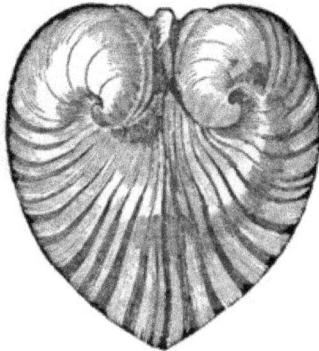

the strength of the cable required to moor it to the rocks or coral reef. The spotted species here figured has a solid and heavy shell, very elegantly shaped, and beautifully marked, as will be seen; the greatly reduced size of the figure prevents anything like justice being done to the original.

The above is a figure of the Heart Isocardea

(*I. cor*), which is also a member of the Clam family, and one of the most elegantly-shaped shells in the whole range of Conchology. It is a native of the Mediterranean and other seas of Europe, and has been taken in deep water on the west coast of Ireland. We complete this group with a representation of the curious *Arcadæ* family, or Ark shells,

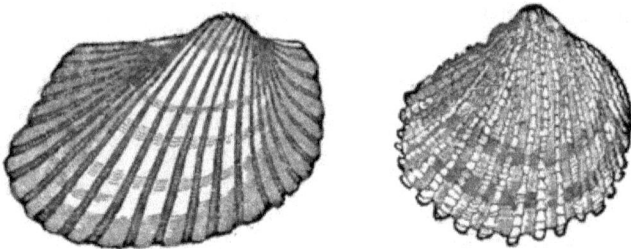

as they are commonly called, because one of the species was thought to resemble the ark built by Noah. Mr. Swainson tells us that the animals of these shells affix themselves to other bodies by a particular muscle, which is protruded through the gaping part of the valves; they also adhere, when young, by means of the bissiform epidermis, or bissus-like outer skin: this species is a native of the Atlantic Ocean and the seas of Europe. The Antique Ark (*A. antiqua*) is very like the Common Cockle, being of a white colour, and heart-shaped.

We here give a representation of this shell, and also of the shell of the pretty little Pearly Trigonia (*T. margaritacea*), included in the *Arcadæ* family; this is a rare species, found only in the seas of New Holland.

SCALLOP SHELLS.

Several species of Scallop Shells are found scattered about on our shores; they belong, as before stated, to the family *Pectinidæ*, the meaning of which term has been already explained. These shells were called by Cuvier "the Butterflies of the Ocean," on account of the various and beautiful colours which they exhibit. Some of them are exceedingly thin, and brittle as glass; one species found in the Arctic regions is as transparent as that substance, and is therefore called *P. vitreus*, from the Latin for glass, which is *vitreum*. One of the commonest of our native Scallops is the St. James's Cockle (*P. Jacobæus*). This shell is found in great plenty along our southern coasts; it is often referred to by old writers, on account of having been commonly worn in the hats of pilgrims to Palestine or the Holy

Land, as the scene of our Saviour's life and death was called. Sir Walter Raleigh, in his poem called "The Pilgrimage," thus enumerates the different articles considered necessary for a Palmer, as these pilgrims were termed :—

> " Give me my *scallop shell* of quiet,
>   My *staff* of faith to lean upon,
>   My *scrip* of joy (immortal diet),
>   My *bottle* of salvation,
>   My *gown* of glory, hope's true gage,
>   And thus I'll make my pilgrimage."

This mollusk, it may be noticed, like many other bivalves, has a flat and a concave or hollow shell. In early times, when plates and drinking vessels were not so plentiful as they are now, one of these served the former purpose, and the other the latter. Thus, in speaking of a feast, a Gaelic or Scottish bard has said—

> " The joy of the *shell* went round."

Sometimes the species termed *Pecten opercularis* was used as the pilgrim's badge : the specific name comes most likely from the Latin *operculum*, whose meaning has been explained.

This, too, is a common British shell, as is also the little speckled Scallop (*P. varia*), which may be found on almost any part of the coast where the water-line is margined with a sandy ridge. The shells are generally about two inches long, of various

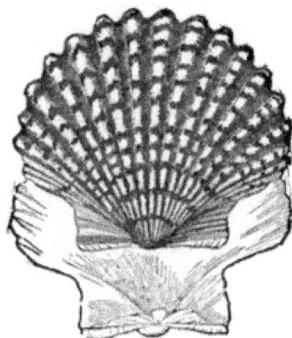

COMMON SCALLOP.

colours, clouded, speckled, and marked with about twelve ribs. There is a foreign species called the Flounder Scallop (*P. pleuronectes*), which is remarkable for having the two valves of the shell of different colours, the upper one being of a rich reddish brown, and the lower one white. The specific name has reference to this, being compounded of the Latin *pleura*, something double, and *necto*, to join. The fish called the Flounder is brown above and white beneath; hence the English name of this shell. The preceding engraving of the Common Scallop, viewed from the front, shows the flat and concave form of the two valves of this shell, and also the depth of the indentations or ridges.

LIMPETS.

Among the rocks of the British coast there are no shells more frequently met with than those of the Common Limpet (*Patella vulgata*); they lie scattered about like so many little empty cups, each having, on the death of the molluks, fallen from the rocky cavity in which it was embedded, and which was just large enough to contain it. Here the animal attaches itself so firmly by its fibrous foot, which is hollow in the centre, and acts like a sucker, that it is almost impossible to loosen its hold otherwise than by inserting something thin, like the blade of a knife, between it and the stone. By this power of adhesion, the Limpet is protected from the violence of the waves, and also from its numerous enemies, aquatic birds and animals, which have a relish for its flesh. Still vast numbers are used as food, both by man and the inferior creatures, so that the means of defence furnished to the Limpets

of the rock are not always sure. "The peasantry
of the western isles of Scotland," we are told by
Miss Pratt, " look to the Periwinkles and Limpets,
which abound on the rocks, for their daily meal,
often for long seasons subsisting almost entirely
upon this humble food. In the Isle of Skye, the
inhabitants are often, at one time of the year, without any other source of provision." Then comes the
Sea-gull, and the Duck, and the Pied Oyster-
catcher, to feed on the poor little mollusk, the bill
of the latter bird being admirably adapted for
loosening its hold on the rock.

Patella in Latin signifies a salad-dish, a knee-
pan, and several other domestic utensils, of a broad
shallow make; and hence we find the plural form
of the word applied to the Limpet family, whose
shells are of such a shape. Members of this family
are found on all rocky coasts, except those of the
Arctic seas; on tropical rocks they grow to a large
size, and form a valuable article of food. A very
curious piece of mechanism is the tongue of the
Common Limpet; it is from two to three inches
long, and has a spoon-like extremity, so that it
looks, when extended from the mouth, like a small
snake. If examined through a microscope, it is
seen to be armed throughout its whole extent with

rows, four deep, of sharp hooked teeth, and between each row are placed two others, which have three points, and are set in a slanting position. The use of this arrangement we cannot at present deter- mine, but no doubt it has a perfect adaptation to the wants of the animal.

There are shell-fish called Key-hole Limpets, which belong to the genus *Fissurellidæ*, from *fissura*, a cleft or slip, from whence comes also fissure. All the members of this genus are distinguished by the aperture at the top of the shell, shaped like a key- hole, which is situated exactly over the breathing organs, and serves as a channel for the water neces- sary for respiration.

Frequently upon the fronds of the large olive sea-weeds may be found a tiny shell shaped some- thing like that of the Common Limpet. It is of an olive-green colour, with blue streaks, and is called, from its clearness, the Pellucid Limpet (*P. pellucida*). There is also another much like it in appearance, which naturalists call *P. lævis*. To the labours of these little mollusks, according to Dr. Harvey, may be attributed the destruction of the gigantic Algæ (sea-weed). Eating into the lower part of the stems, and destroying the branches of the roots, they so far weaken the base that it becomes unable

to support the weight of the frond, and thus the
plant is detached and driven on shore by the waves.

> "And so the forest tall, that groweth
>   Underneath the waters clear,
> Does the little woodman mollusk
>   Level every year;
> From small causes, great results—
>   Teaching you to persevere."

ROCK-BORERS.

The family *Pholadæ* comprises a group of mollusks, the boring habits of which have long been
known; they penetrate wood, hard clay, chalk, and
rocks, and devastate the labours of man; they
attack the hulls of ships, the piles which form the
foundations of piers and breakwaters, and they force
themselves upon our attention by the loss of property, as well as life, which results from their
hidden depredations. Of this family, those belonging to the genus *Pholus* may be more especially
likened to the Edomites of Scripture, because they
take up their abode in the rock, and hollow out for
themselves dwellings therein. With a shell as thin
as paper, and brittle as glass, the wonder is how

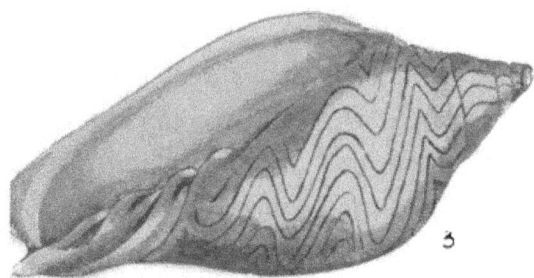

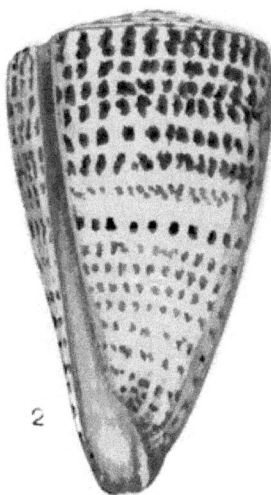

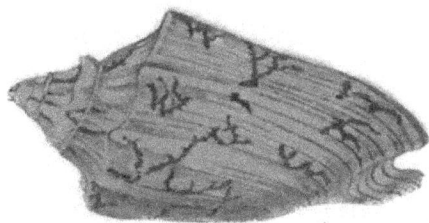

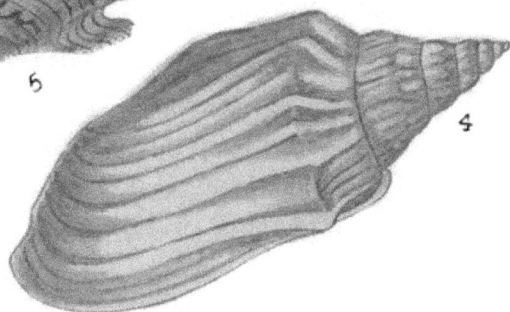

3

2

5

4

Plate V

these Rock-borers work their way into and through hard stones. Some naturalists assert that they effect this by means of an acid which decomposes the substance of the rock, and renders it soft; others, that the animal keeps turning round and round like an instrument called an auger, and so gradually rasps away the surface of the stone with the angles of its shell, but we question whether the shell would not be worn out first in such a process. The generic name of these " stone-piercers," comes from the Greek word *Pholeo*—to hide, and the rocky chambers which they hollow out for themselves, are as snug hiding-places as can well be imagined; yet however deep they may go into these gloomy caverns, as we should be apt to suppose them, they need not be in darkness; for it appears that these Pholades emit a most remarkable light—whether phosphorescent or not does not appear to be determined; so strong is it, that it is said to illuminate the mouth of the person who eats the mollusk; and it is remarked by Dr. Priestly, that " contrary to the nature of most fish, which give light when they tend to putrescence, this is more luminous the fresher it is, and when dried its light will revive on being moistened with water." So that in more respects than one these Rock-borers are mysteries. The most common of them, perhaps, is the Prickly Pidduck, or Peckstone (*P. dactylus*), which is much used by the fishermen of our coasts as bait; the specific name is the Latin for a fruit shaped like a finger, which is something like the shape of this mollusk, as will be seen by the annexed engraving.

The genus *Pholus* is very widely distributed, and all the species have the same boring habits as those of our own coast, which we need not enumerate. Like them, too, in this respect, are the marine worms called *Teredo*, which make their way into the bottoms of ships, and all submerged timber, but these will be more fully spoken of in another volume. The above figure exhibits the *Pholas dactylus* as it appears in a section of rock, split open for the purpose of seeing the shelly miner at his work.

MULTIVALVES.

WE have insensibly passed from the Bivalve shells to those composed of several pieces, and therefore called Multivalves; probably, perhaps, the Rock-borers, last described, come into this division, for although their covering consists mainly of two principal portions or valves, yet there are often additional parts; in some a calcareous tube enve-lops the whole mollusk, leaving only an opening behind. This is more especially the case with those which most resemble worms, such as the genera *Teredina* and *Teredo,* included by Lamarck in the family which he calls *Tubulidæ.*

The first group of multivalves we shall have to notice, are

THE CHITONS,

forming the family *Chitonidæ.* The term has a Greek derivation, and means a coat of mail. These mollusks are covered by a shell formed of eight distinct portions, arranged along the back in a single row, and attached to a mantle which re-

resembles leather, being very tough and wrinkled; the edges of this mantle extend beyond the borders of the plates, which overlap each other, so as to constitute a kind of armour, very different from the conical shell of the Limpet, or the turbinated, that is twisted, case of some of the Borers. The coverings of the Chitons are variously marked, so that each distinct species is known by its peculiar pattern, as a knight of old by the quarterings of his shield. All the mantles, however, have scaly, hairy, or spiny margins. In this coat of mail the animal can roll itself up like an armadillo, and so be tolerably secure from its enemies; it has an oval foot, the sides of which are covered with small leaflets, and by means of this it can attach itself to rocks, like the Limpet, or travel about in search of adventures. It has no distinct head, therefore it is *acephalous;* nor any perceptible eyes. The mouth is furnished with a long tongue, curled up spirally, like a watch-spring, and armed with horny teeth.

The members of the Chiton family are numerous, being found on most rocky shores; they attain the largest size in the hottest climates, having never been found very far north. The British species are small, and not more than two or three in number; they may be found adhering to stones near low-water mark. We give a figure of one of these, called the Tufted Chiton (*C. fascicularis*). This word is from the Latin *fasciculus*, a little bundle of leaves or flowers, and it refers to the hairy tufts that edge the mantle of this marine slug.

BARNACLES,

Or, as they are sometimes called, Bernicles, belong to what naturalists term the class *Cirrhopoda*, sometimes spelled *cirripeda*, which appears to be derived from the Latin *cirrus*—a tuft or lock of hair curled, and *pede*—a foot; hence the term may be translated hairy-footed. Such of our readers as have seen the Common or Duck Barnacle (*Pentalasmis anatifera*) will at once understand the applicability of this term. Many a piece of drift-wood comes to land literally covered with long fleshy stalks, generally of a purplish red colour, twisting and curling in all directions, and terminating in delicate porcelain-like shells, clear and brittle, of a white colour, just tinged with blue, from between which project the many-jointed *cirrhi*, or hair-like tentacles, which serve the purpose of a casting-net, to seize and drag to the mouth of the animal its prey, which consists of small mollusks and crustacea.

This is the Barnacle about which such strange stories are told by old writers, who affirmed that the Barnacle or Brent Goose, that in winter visits our shores, is produced from these fleshy foot-stalks and hairy shells by a natural process of growth, or, as some philosophers of our day would say, of development. Gerard, who, in 1597, wrote a "Historie of Plants," describes the process by which the fish is transformed into the bird; telling his readers that as "the shells gape, the legs hang out, that the bird growing bigger and bigger, the shells open more and more, till at length it is attached only by the bill, soon after which it drops into the sea; there it acquires feathers, and grows to a fowle." There is an amusing illustration given in Gerard's book, where the young Geese are represented hanging on the branches of trees, just ready to drop into the water, where a number of those that have previously fallen, like ripe fruit, and attained their full plumage, are sailing about very contentedly. It was part of this theory that the Barnacles were of vegetable origin, they grew upon trees, or sprung out of the ground like mushrooms; so we find, in the works of an old poet named Du Bartas, these lines :—

" So slow Bootes underneath him sees
   In the icy islands goslings hatched of trees,
   Whose fruitful leaves, falling into the water,
   Are turned, as known, to living fowls soon after;
   So rotten planks of broken ships do change
   To Barnacles. O transformation strange !
   'Twas first a green tree, then a broken hull,
   Lately a mushroom, now a flying gull."

The investigations of modern science have quite exploded this foolish notion; we now know exactly what transformations the Barnacle undergoes; strange enough some of them are, but it does not change into a Goose, although its specific name has reference to that bird, being derived from *anas,* the Latin for Goose.

The shell of the Barnacle is composed of five pieces joined together by membranes; four pieces are lateral, that is to say, they form the sides, the word comes from the Latin *latus*—a side; the other is a single narrow slip, which fills what would otherwise be an open space down the back between the valves; these parts of the shell appear to be somewhat loosely connected, so as to allow free action to the animal lodged within, which is enclosed in a fine skin or mantle. The mouth is placed at the lower part, near the opening, whence the *cirrhi* issue forth ; this mouth is a curious piece of mechanism, being furnished with a horny lip covered with minute *palpi*, or feelers ; there are three pairs of *mandibles*, that is jaws, the two outer ones being horny and *serrated*, that is jagged or toothed like a saw ; the inner one is soft and membranous, that is, composed of little fibres, like strings, crossing each other, as we see what are called the veins in a leaf.

Much more might be said about the internal structure of the Cirrhopods, or *Balani*, as the Barnacle group is sometimes called, from the Latin *Balanus*—a kind of acorn. By some naturalists, the term is not applied to the stalked *Cirrhipoda*, like that we have been describing, but only to the *sessile* kinds, that is, those which set close or grow low ; from the same Latin root comes the English word *session*—a settling. The coverings of these Dwarf Barnacles are sometimes called acorn shells ; they are commonly white, of an irregular cone shape, composed of several ribbed pieces, closely fitted together with an opening at the top, closed by an *operculum*, or stopper.

These shells cover in patches the surface of

exposed rocks, drift wood, and any other substance. Some of the mollusks affix themselves to the bodies of whales, others form a lodgment in the hollows of corals and sponges. Once fixed they remain so during life, taking their chance of such suitable food as may come within their limited sphere of action. At an earlier stage of their existence, both their shape and habits are very different, being lively little creatures, swimming about hither and thither like water-fleas. They are about the tenth of an inch long, and of most grotesque appearance, having six-jointed legs set with hairs, the whole being so arranged that they act in concert, and striking or flapping the water, send the little body along in a series of bounds; then the creature has two long arms, each furnished with hooks and a sucker, and a tail tipped with bristles, which is usually folded up under the body; its pair of large staring eyes are *pedunculated*, that is, set upon foot-stalks; it has a house on its back, like a bivalve shell, into which it can collect its scattered members when occasion requires. When of sufficient age to settle itself in life, and become a staid member of submarine society, it fixes itself to some convenient object, throws away its eyes as no longer useful, gets rid of its preposterous limbs, enlarges its house, and sits down to fishing in a small way for an honest and respectable livelihood.

A piece of timber covered with Stalked Barnacles, wriggling and twisting about like so many helmeted snakes, and waving their plume-like *cirrhi,* is a very curious sight.   They sometimes

attach themselves to ships' bottoms in such numbers as to retard their progress through the water; they do not, however, bore into and destroy the timber, like the *Teredines,* or ship-worms, to which we have alluded in our brief notice of the *Pholadœ.* The growth of Barnacles must be very rapid, as a.

ship perfectly free from them, will often return after a short voyage, with her bottom below the water-line completely covered.

We give a representation of a group of these stalked mollusks, as they appear affixed to a piece of timber. This is the Common, or Duck Barnacle.

CUTTLE-FISH.

Strange and monstrous as are the forms of many of the creatures that inhabit the deep, there are, perhaps, none more so than those belonging to .that division of the class *Cephalopoda,* called *Sepia,* or Cuttle-fish. But before we go any further, let us inquire what is meant by a *Cephalopod.* Our readers have already learned that *Gasteropod* means stomach and foot, and that *acephalous* means headless; now here we have a word which takes a portion of each of the others (*cephal*—head, and *peda,* or *poda*—a foot), consequently *ceph-a-lo-po-da* is a class of molluscous animals which have their feet, or organs of motion, arranged round the head, something, you may suppose, like the celebrated hero of nursery rhymes,

"Tom Toddy, all head and no body."

Only our bag-shaped Mr. Sepia, with his great,

round, staring eyes, and numerous legs or arms, whichever you please to call them, all twisting and twining about like so many serpents, is a much more formidable-looking individual. A strange fellow is this altogether; he has a shell, but he does not use it for a covering, he carries it inside of him, and it serves the purpose of a sort of backbone. It is horny and calcareous, light and porous, as our readers well know, having most likely often used it to take out unsightly blots, or erase mistakes in their copy or cyphering books.

When Mr. Sepia walks abroad, he sticks his little round body upright, so that his eyes and mouth, which is armed with a parrot-like beak, are brought close to the surface over which he passes, while his long twining legs go sprawling about in all directions. On the insides of these legs are a great number of small circular suckers, by means of which the animal can fix itself to any object so tightly, that it is almost impossible to detach it without tearing off part of the limb. Woe be to the poor unfortunate fish that chances to come in its way; the snaky arms are thrown around it, and made fast, and away goes the cephalopod for a ride, eating on the road to lose no time, on the finny steed that carries it. In some species each of the suckers has a hook in the centre, which, of course, renders the hold yet firmer, and, no doubt, adds to the disagreeable sensation which their tight compression must cause. It is likely that these hooks are intended to retain the hold of soft and slippery prey, which might otherwise be too agile for the "ugly customer," that would affectionately embrace it. But Mr. Sepia, though well armed in front, is rather open to attacks in the rear of his soft naked body. To provide for such an emergency, he is furnished with a little bag of inky fluid, which he squirts out in the face of his pursuer, and escapes under cover of the cloud. This is the substance used by painters, and called sepia, whence the generic name of the mollusks which produce it.

In the British seas none of these cephalopods attain so large a size as to be formidable to man, as they do in warmer climates. It was asserted by Dens, an old navigator, that in the African seas, while three of his men were employed during a calm in scraping the sides of the vessel, they were attacked by a monstrous Cuttle-fish, which seized them in its arms, and drew two of them under water, the third man was with difficulty rescued by cutting off one of the creature's limbs, which was as thick at the base as the fore-

yard of the ship, and had suckers as large as
ladles; the rescued sailor was so horrified by the
monster that he died delirious a few hours after.
An account is also given of another crew who
were similarly attacked off the coast of Angola;
the creature threw its arms across the vessel, and
had nearly succeeded in dragging it down, and
was only prevented doing so by the severing of
its limbs with swords and hatchets. A diligent
observer of nature has asserted that in the Indian
seas Cuttle-fish are often seen two fathoms broad
across the centre, with arms nine fathoms long.
Only think, what a monster! with a body twelve
feet across, and eight or ten legs like water-snakes,
some six-and-thirty feet long. Well may it be said,
that the Indians when they go out in boats are in
dread of such, and never sail without an axe for
their protection.

There is a story told by a gentleman named
Beale, who, while searching for shells upon the
rocks of the Bonin Islands, encountered a species
of Cuttle-fish called by the whalers "the Rock-
squid," and rashly endeavoured to secure it. This
cephalopod, whose body was not bigger than a
large clenched hand, had tentacles at least four feet
across, and having its retreat to the sea cut off by

Mr. Beale, twined its limbs around that gentleman's arm, which was bared to the shoulder for the
purpose of thrusting into holes of the rocks after
shells, and endeavoured to get its horny beak in a
position for biting. The narrator describes the
sickening sensation of horror which chilled his very
blood, as he felt the creature's cold slimy grasp,
and saw its large staring eyes fixed on him, and the
beak opening and closing. He called loudly for
help, and was soon joined by his companion, who
relieved him by destroying the Cuttle-fish with a
knife, and detaching the limbs piece by piece.

There are several species of these cephalopods;
the most generally distributed appears to be the

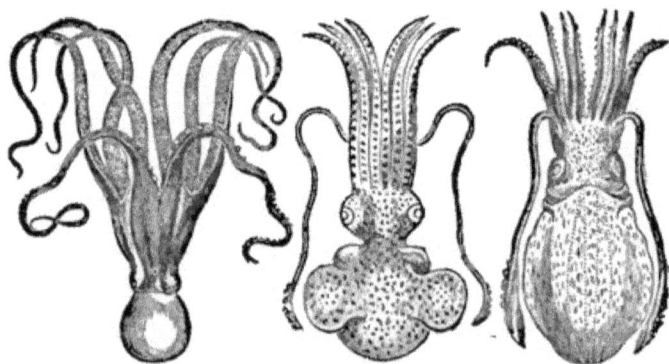

O. VULGARIS. S. VULGARIS. S. OFFICINALIS.

Octopus vulgaris, or Common Cuttle-fish, which is

sometimes found on our own shores, where also **may** be obtained the Common Sepiola, *S. vulgaris*, usually about three inches long, and the Officinal Cuttle-fish, *S. officinalis*, which is about a foot in length; we give below small figures of each of these three species, to show the difference in the shape: the two last, it will be observed, have, in addition to the eight tentacles, which give the generic name *Octopus*, signifying eight, two long side arms, the use of which does not appear to be very clearly determined.

NAUTILUS AND AMMONITE.

The Nautili are called testaceous cephalopods; our readers know, or ought to know, the meaning of both these terms.   Like the Cuttle-fish, they are sometimes called *Polypi*, because they have many arms or tentacles, the word *poly*, with which a great number of English words commence, being the Greek for many.   An ancient writer named Aristotle, after describing the naked cephalopods, says, "There are also two polypi in shells; one is called by some, *nautilus*, and by others, *nauticus*.   It is like the polypus, but its shell resembles a hollow

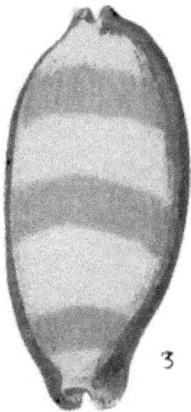

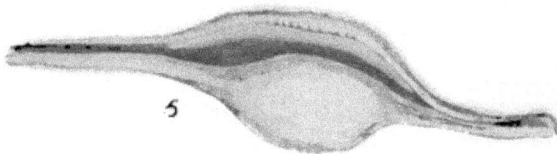

Plate VII

comb or pecten, and is not attached. This polypus ordinarily feeds near the sea-shore; sometimes it is thrown by the waves on the dry land, and the shell falling from it, is caught, and there dies. The other is in a shell like a snail, and this does not go out of its shell, but remains in it like a snail, and sometimes stretches forth its *cirrhi*." The first of these animals, there can be no doubt, is the Argonaut, or Paper Nautilus, and the latter that which is called the True Nautilus, of both of which species let us say a few words, which we will introduce by quoting some beautiful lines from a poem called " The Pelican Island," by James Montgomery :—

> " Light as a flake of foam upon the wind,
>   Keel upwards from the deep, emerged a shell,
>   Shaped like the moon ere half her orb is filled:
>   Fraught with young life it righted as it rose,
>   And moved at will along the yielding water.
>   The native pilot of this little bark
>   Put out a tier of oars on either side;
>   Spread to the wafted breeze a two-fold sail,
>   And mounted up and glided down the billow,
>   In happy freedom, pleased to fill the air,
>   And wander in the luxury of light."

The tiny mariner here alluded to is the Paper Nautilus, common in the Mediterranean and some

Q

tropical seas; its scientific name is *Argonauta argo*.
In the mythology, we read that *Argo* was the name
of a ship that carried a certain Grecian named
Jason, and a crew of *argives*, in search of adventures. Some say that the term is derived from a
Greek word signifying swift. This party of mari-
ners, said to be the first that ever sailed upon the
sea, was called *Argonauts*, or, as it might be freely
translated, seamen of the ship Argo. *Nauticus*, in
Latin, signifies anything relating to ships or navigation, and here you have the whole origin of the
name of this little Argonaut, about which we must
sing you a song written by Mary Howitt before we
proceed further :—

" Who was the first sailor ? tell me who can ;
    Old father Neptune !—no, you're wrong,
There was another ere Neptune began ;
    Who was he ? tell me. Tightly and strong,
      Over the waters he went—he went,
      Over the waters he went !

" Who was the first sailor ? tell me who can ;
    Old father Noah !—no, you're wrong,
There was another ere Noah began,
    Who was he ? tell me. Tightly and strong,
      Over the waters he went—he went,
      Over the waters he went.

' Who was the first sailor? tell me who can;
 Old father Jason!—no, you're wrong,
There was another ere Jason began,
  Don't be a blockhead, boy! Tightly and strong,
   Over the waters he went—he went,
   Over the waters he went.

" Ha! 'tis nought but the poor little Nautilus—
 Sailing away in his pearly shell;
He has no need of a compass like us,
 Foul or fair weather he manages well!
  Over the water he goes—he goes,
  Over the water he goes."

Many more poems of the like nature we might quote, for this little shelled cephalopod has been a favourite with the poets time out of mind, and in some instances they and the less imaginative naturalists have disagreed in their accounts of its form and operations; for instance, Pope says—

" Learn of the little Nautilus to sail,
 Spread the thin oar and catch the driving gale."

" Catch a fiddle-stick," say some naturalists, the little Nautilus does nothing of the sort; and if you go to him to learn navigation, you will never be much of a sailor. He may teach you how to sink to the bottom and rise again, and that kind of knowledge might be worth something to you if you could breathe under water; and he might teach you how to swim, but not how to sail, for, in spite of all poetic theories, he does the former and not the latter. Most usually he walks about at the bottom of the sea on his long arms, something like the Cuttle-fish, feeding on the marine vegetation; the shell is then uppermost. If we could look inside of it we should see numerous little chambers or cells, the larger and outermost of which only are inhabited by the mollusk, the others being filled with air render the whole light and buoyant. Through the centre of these chambers, down to the smallest of them, runs a membranous tube which can be exhausted or filled with fluid at the pleasure of the animal, and the difference thus effected in the weight of the shell enables it to sink or swim; in the latter case, up it goes to the surface, and "keel upwards from the deep," emerges, as the poet has said, but once there it soon reverses its position. The shell becomes like a boat it is true, but its inhabitant neither points a sail nor plies the oar, but propels itself along stem foremost by a muscular action, which by alternately compressing and loosening a kind of siphon, throws out jets or gushes of water, which, by the resistance they meet with from the surrounding fluid, give the desired onward motion, and away the swimmer goes, his long arms gathered closely together, and streaming behind like the tail of a comet, and its round eyes keeping a sharp look-out on either side. Should it espy danger, the body and limbs are withdrawn into the shell, and the fluid driven through the central tube, so as to compress the air in the pearly cells, and down sinks the swimmer once again to his native depths, where

> "The floor is of sand like the mountain drift,
>   And the pearl shells spangle the flinty snow;
> And from coral rocks the sea-plants lift
>   Their boughs where the tides and billows flow.
>   The water is calm and still below,
> For the winds and waves are absent there;
>   And the sands are bright as the stars that glow
> In the motionless fields of upper air.
> And life in rare and beautiful forms
>   Is sporting amid those bowers of stone,
> And is safe, when the wrathful spirit of storms
>   Has made the top of the waves his own."

On the following page we give two figures of the Argonaut, one of which represents him crawling at the bottom of the sea, and the other swimming on the surface.

The True, or Pearly Nautilus (*N. Pompilius*),
the origin of whose specific name we have been
unable to discover, is much like the Argonaut in
appearance and general construction ; the shell is
externally smoother and more iridescent, it is also
generally somewhat thicker than the former kind,
and has internally more chambers or divisions ; its
pearly lustre renders it a beautiful ornament, and
the large size it frequently attains a very con-

spicuous one. Its inhabitant has several peculiarities of organization, which distinguish it from
the Argonauts, but into these we need not enter ;
neither can we pause to describe the other species
of Nautili, the shells of which, like those of the
Cowry and other univalves, are covered with a
membrane which hides their beauty. This membrane or mantle sometimes extends some distance
beyond the edge of the shell, and, being of a light
and filmy appearance, may have been mistaken for
a sail hoisted by the creature to catch the breeze,

while its long arms, thrust up into the air or down into the water, may have been thought to be masts or oars, so that the poets are not so much to be blamed, if they say as Wordsworth does—

> " Spread, tiny Nautilus, the living sail
> Dive at thy choice, or catch the freshening gale."

Nearly allied to the Nautili are these beautiful fossil shells called Ammonites, from their fancied resemblance to the horns of a heathen deity or god, called Jupiter Ammon. These shells, at once the wonder and pride of geologists, are found in the chalk formations, and thousands of years must have passed away since they were inhabited by living creatures. The Nautili which swam and sported with them at the depths of the ocean, as is proved by the shells of many species found in the same chalky deposits, have still their living representatives; but those winding galleries and pearly chambers once fragile as paper and brittle as glass, now turned into, and surrounded by solid stone are all shells of extinct species, and we can hardly see and handle them without some degree of awe and reverence; when we reflect on the great and wonderful changes that have passed over the earth since they were formed by a hand Divine, instinct

with the breath of life, and then to be embedded in
the rock as everlasting characters by which the
unborn generations of men might read in history
of those changes, and of the providential dealings
of God with his creatures. Of these Ammonites,
and other fossil shells, much more will have to be
said in our proposed geological volume; the poem
which follows will very appropriately conclude the
above remarks, and our present little work on

 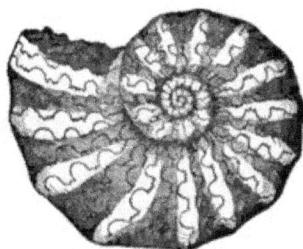

NAUTILUS.   AMMONITE.

shells—beautiful, wonderful shells! useful, ornamental, instructive! The subject is one which we
would earnestly invite our young readers to study:
it is but here introduced; we have picked up a few,
very few, of the wonders and beauties of conchology, and presented them to their notice in the
hope that they may be induced to desire a more
intimate acquaintance with this branch of natural

science, which has been hitherto greatly neglected. To understand it thoroughly, much attention and perseverance will be required, but even a slight acquaintance with it will yield both pleasure and profit to the mind.

THE NAUTILUS AND THE AMMONITE.

The Nautilus and the Ammonite,
  Were launched in storm and strife;
Each sent to float in its tiny boat,
  On the wide, wild sea of life.

And each could swim on the ocean's brim,
  And anon its sails could furl,
And sink to sleep in the great sea deep,
  In a palace all of pearl.

And theirs was a bliss more fair than this,
  That we feel in our colder time;
For they were rife in a tropic life
  In a brighter, happier clime.

They swam 'mid isles whose summer smiles
  No wintry winds annoy;
Whose groves were palm, whose air was balm,
  Whose life was only joy.

They roam'd all day through creek and bay,
  And travers'd the ocean deep;
And at night they sank on a coral bank,
  In its fairy bowers to sleep.

And the monsters vast of ages past,
  They beheld in their ocean caves;
And saw them ride in their power and pride,
  And sink in their billowy graves.

Thus hand in hand, from strand to strand,
  They sail'd in mirth and glee;
Those fairy shells, with their crystal cells,
  Twin creatures of the sea.

But they came at last to a sea long past,
  And as they reach'd its shore,
The Almighty's breath spake out in death,
  And the Ammonite liv'd no more.

And the Nautilus now in its shelly prow,
  As o'er the deep it strays,
Still seems to seek, in bay and creek,
  Its companion of other days.

And thus do we, in life's stormy sea,
  As we roam from shore to shore;
While tempest-tost, seek the lov'd—the lost,
  But find them on earth no more!

G. F. RICHARDSON.

SHELLS OF COMMON OCCURRENCE.

BY WILLIAM WALLACE FYFE.

The *Limacidæ*, or land-slugs, are represented con-
chologically by the thin crustaceous shells found on
dissection within their mantles, being, as every-
body familiarly knows, outwardly destitute of shell.
This shield is protective of the cavity employed in
respiration. Figs. 1, 2, 3, and 4 represent four of
these shields, extracted from the milky, yellow,
tree, and spotted slugs respectively (*Limax agrestis,
L. flavus, L. arborum, L. cinereus*).

These creatures, as every lover of a garden too
well knows, are powerful vegetable feeders, making
their appearance in damp weather in multitudes
like an Egyptian plague. Their destructive voracity
enables them to secrete an exuberance of white
milky mucilage from their bodies, to discharge this
copiously when irritated, and to mark their devouring tracks in their slime. Like linseed and other

mucilaginous matter, animal and vegetable, slugs, when boiled, have been employed as a cure for consumption. When hard pressed by hunger only will they eat dead earth-worms, and hence their blight falls chiefly on the growing plant. The observer may occasionally have felt startled to see the *Limax* suspended by an almost invisible but very tenacious thread which it possesses the power of spinning, betwixt him and the light. This is

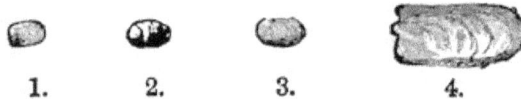

1. Limax agrestis (the Milky Slug), *Müller.* 2. L. flavus (the Yellow Slug), *Linnæus.* 3. L. arborum (the Tree Slug), *Chautereaux.* 4. L. cinereus (the Spotted Slug), *Müller.*

used by the slug to drop from on high. Like the spider, it exudes this mucous thread from the secretions of its body. Encumbered with no mansion which it must carry on its back like the snail (*Helix*), the slug is yet more hardy without its shelter than the *Helix*, and remains active far into the winter, when the other lies dormant in the crevice of the wall.

The most common slug of the fields, *L. agrestis,* or milky slug, about an inch and a half long, is the

most destructive of molluscous animals, devouring
the roots of plants as well as their leaves. It is
bisexual and very prolific, breeding several times
a year; and Chautereaux counted three hundred
and eighty eggs deposited by two individuals betwixt April and November, laying from thirty to
seventy at a time. They rapidly increase in size,
and reach maturity in three months, although they
probably begin to lay eggs in about two. The *L.
flavus*, or *variegatus* (for although the lower surface
be yellowish-grey, the upper is thickly and irre-
gularly-spotted blackish brown), has a limpid slime,
but secretes, when irritated, a thicker bluish-white
mucus. On reaching any twig or place where there
occurs a difficulty of proceeding, it allows itself to
drop or be slowly lowered by its thread of adhesive
mucus, which at length gives way. This is the
slug which is found under damp turf and stones,
beside walls and among plants. It is twice the size
of the field slug, being from three to four and
sometimes even five inches long. Mr. E. J. Lowe
says it is best known as the cellar slug, and is
in such situations gregarious. The tree slug (*L.
arborum*) is greyish, with marbled side-stripes, and
a dusky band along the back. It feeds upon wood
and affects decaying trees, whence it uses its

mucous thread to descend. The spotted grey slug (*L. cinereus*), best known as the black slug, but not to be confounded with the *Arion*, though less common than the field slug, is very abundant in damp situations, under decaying wood and fragments of stones in gardens, hedges, etc., and in cellars and outhouses; copious rain, or even dew, enticing it from its retreat. It is the largest of the slug family, being six inches long, and will be found handsomely formed when attentively surveyed. It is circular on the back, acute and pointed at the tail; with upper tentacles of great length and short lower ones; it secretes colourless mucus. The shield is slightly stained with pink. It deposits during spring from fifty to sixty eggs, attached in heaps together, under stones and at the roots of grass and trunks of trees.

The *Helix*, or snail, has a shell spirally rolled, and although possessing no operculum, it substitutes that filmy mucous covering, by means of which it closes up its shell, and which is perforated with holes to enable it to breathe. Remaining concealed in obscurity during the day, it comes forth to feed evening and morning, or after rain, retiring from business altogether in the winter into a hole or crevice, or amongst moss, and shutting

up shop till the return of solar heat. When in
motion it carries its shell balanced obliquely on its
back, and keeps advancing and retracting its tentacles. It is quite as destructive to vegetation as
the slug.

We should have enough to do to describe all
the species and varieties, for of Helices alone we

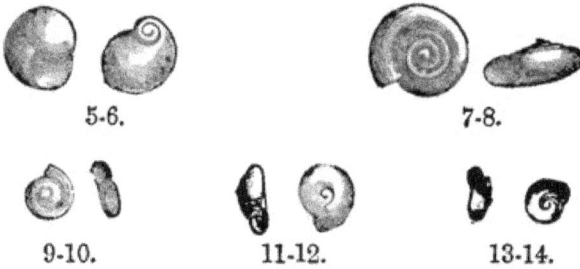

5-6.　　　　　　　　7-8.

9-10.　　　11-12.　　　13-14.

5-6. Vitrina pellucida (the Transparent Glass Bubble Shell),
　　Müller. 7-8. Zonites cellarius (the Cellar Snail), *ibid.*
　　9-10. Z. alliarius (the Garlic Snail), *ibid.* 11-12. Z.
　　nitidulus (the Dull Snail), *Draparnaud.* 13-14. Z. purus
　　(the Delicate Snail), *Alder.*

may count at least forty. The *Vitrina pellucida,*
or transparent glassy snail (Figs. 5 and 6), found
among the putrescent leaves, moss, and decaying
wood of plantations and hedge-bottoms, and also
under stones, is an exceedingly thin, transparent,
glossy, smooth, and fragile watery-green shell,
through which the mantle of the animal within is
reflected. It is elliptical in form, with three and a

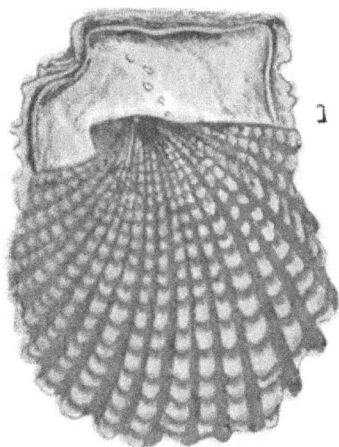

1

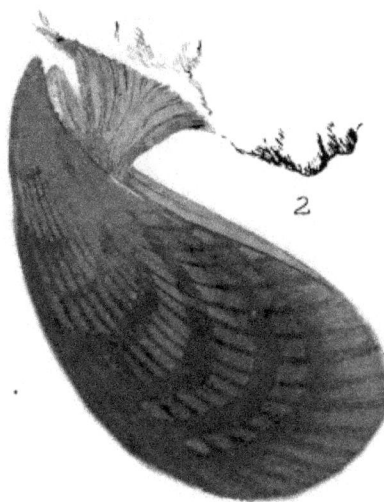

2

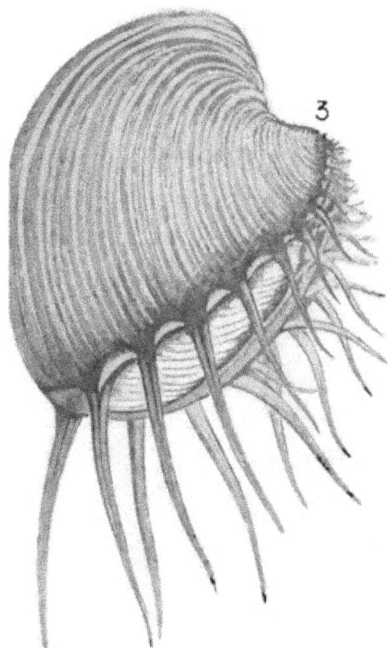

3

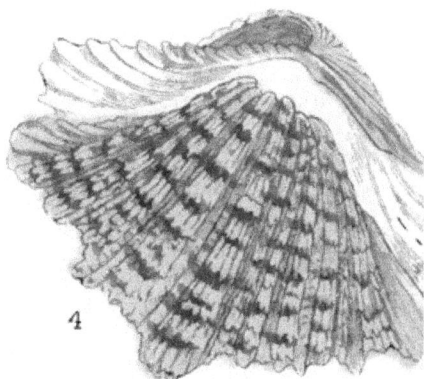

4

half depressed whorls. The shell of the *Zonites cellarius*, a cellar snail (Figs. 7 and 8), is also shining, smooth, and pellucid, and of a pale yellowish horn-colour. It is found in cellars, drains, and shady courts, in fields and woods, under stones, and amongst grass. The shell of the garlic snail (*Zonites alliarius*, Figs. 9 and 10), is nearly flat, and more convex, yellower in colour, but equally pellucid, smooth, polished, and fragile. Some of these creatures have, when alive, a strong odour of garlic, some have it on being plunged in hot water (which is the readiest way of killing them for the shell), though not when alive. Its numbers in our bag, as swept down a river, are somewhat extraordinary.

The little *Zonites nitidulus* (Figs. 11 and 12), takes from its shell the name of little shining snail. A deep umbilicus is seen in the shell. The animal is also called the " dull snail," from its leaden colour; but the shell, three-tenths of an inch in diameter, is of a yellowish horn-colour, and very like *Z. cellarius*. Another of these small shells, the delicate snail, *Z. purus* (Figs. 13 and 14), is only two lines or less in diameter ; it is not very common, but, like the rest, smooth, glossy, and transparent, and may be known by its mouth, which is placed obliquely. It cannot well be confounded with *Z. radiatus,* the rayed snail shell (Figs. 15 and 16), for that, although polished, shining, and pellucid, is regularly striated or wrinkled, and is horn or amber coloured, and two lines in diameter. *Z. excavatus* (Figs. 17 and 18) is a quarter of an inch shell, of which there are multitudes in our bag, found under felled timber

15-16.   17-18.

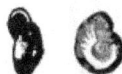

19-20.   21-22.

15-16. Z. radiatus (the Rayed Snail), *Alder.* 17-18. Z. excavatus (the Excavated Snail), *Bean.* 19-20. Z. nitidus (the Shining Snail), *Müller.* 21-22. Z. crystallinus (the Crystalline Snail), *ibid.*

and decayed wood. *Z. nitidus,* the shining snail (Figs. 19 and 20) is likewise pellucid, the colour being brownish horn, its diameter a quarter of an inch. In pine-beds and damp hothouses, where it is found largely congregated, sad havoc is perpetrated by this tiny snail. The *Z. crystallinus* of Müller (Figs. 21 and 22) scarcely exceeds an eighth of an inch in diameter, and occurs amongst decayed leaves and stones.

Of the Helices proper, we figure fifteen different species; they are mostly too well known to require to be more than mentioned. Of these the largest and most familiar, *Helix aspersa,* or common snail (Fig. 23), is the largest and most destructive in the

23.   24-25.

26-27.

3. Helix aspersa (the Common Snail), *Müller.* 24-25. H. revelata (the Green Snail), *Férussac.* 26-27. H. nemoralis (the Girdled Snail), *Linnæus.*

garden. Its usual diameter is an inch and a-half. It is olive coloured, with dark brown bands. The shell is in reality not slimy, though apparently rough on the surface. It is alleged that barrels are exported as a dainty to America, and that the London markets are largely supplied with it as a remedy for pulmonary complaints. In the United States it has been successfully acclimatized, and is now getting common. It has a penchant for nettles, wild celery, elder *Primula vulgaris*, and will climb walls, and apple and scented poplar trees, to a great height, but is capable of a long fast. Mr. Lowe mentions one that fasted one hundred and eight days in summer. The green snail (*H. revelata*, Figs. 24 and 25) is very rare and pretty. It was added by Edward Forbes in 1839; he found it near Doyle's monument in Guernsey. The yellowish-green *H. nemoralis*, or girdled snail (Figs. 26 and 27), is abundant and beautiful, and known to every

28-29.

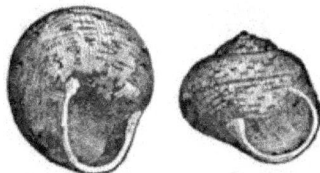

30-31.

28-29. H. hortensis (the Garden Snail), *Montagu.*   30-31. H. arbustorum (the Shrub Snail), *Linnæus.*

one. This snail also has been introduced into North America, where it is becoming common. It is about seven-eighths of an inch in diameter. In about an hour after a shower, the banks they frequent become quite covered over with them, where in dry weather not one is to be found, as they retire into holes in the ground, and amongst grass, roots, and rubbish. The *H. hortensis*, garden snail (Figs. 28 and 29), is, like the preceding, very varied in its colours, though less in its size, which is three-fourths of an inch in diameter. The shrub snail (Figs. 30 and 31), which nearly resembles this, is pretty and interesting. The zoned snail *H. virgata* (Figs. 32 and 33), is more peculiar to

32-33.   34-35.

36-37.   38-39.

32-33. H. virgata (the Zoned Snail), *Da Costa.*   34-35. H. caperata (the Black-tipped Snail), *Montagu.*   36-37. H. ericetorum (the Heath Snail), *Müller.*   38-39. H. hispida (the Bristly Snail), *Linnæus.*

chalk and lime districts; and the little black-tipped snail, *H. caperata* (Figs. 34 and 35), which might at first sight be mistaken for the zoned snail, is a

Dorsetshire snail, added by Dr. Pulteney. Of the others which are figured, the heath snail is pale green in colour, and the bristly snail (*H. hispida*), which associates with it on the dry heath, is nearly greyish. The last shell is scattered over with bristles, but its diameter is only a quarter of an inch. The prickly snail (Fig. 42), though more

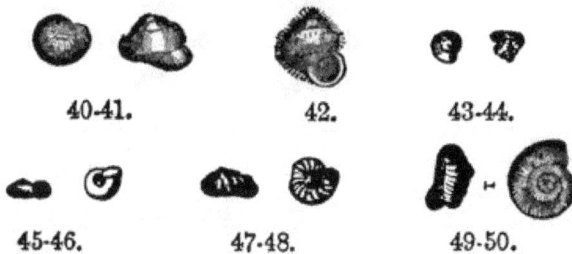

40-41.   42.   43-44.

45-46.   47-48.   49-50.

40-41. H. sericea, *Draparnaud.* 42. H. aculeata (the Prickly Snail), *Müller.* 43-44. H. fulva (the Top-shaped Snail. 45-46. H. pulchella (the White Snail), *Müller.* 47-48. H. rotundata (the Radiated Snail), *ibid.* 49-50. H. pygmœa (the Pigmy Snail), *Draparnaud* (much exaggerated, see scale betwixt).

minute, is still more remarkable from having the appearance of large prickles on its shell. Amongst these very small shells are the top-shaped and white snail, the radiated, and finally the pigmy snail, whose diameter is less than a line.

We shall still go to our bag for one more handful of miscellaneous shells, which differ much from the Helices now disposed of so far as we have got them in our budget. Our greatest regret is the absence of *H. pomatia,* the largest of the British land shells, the favourite shell food of the Romans, still eaten in many parts of modern Europe.

The *Bulimus obscurus,* dull or dusky twist shell, is considered to derive its first or generic name from a Greek term signifying insatiable hunger (βούλιμος), and its specific from the Latin, indicative of its colour (dusky or dull), for scientific jargon is generally macaronic, and by no means minds a Babel of languages. The shell is not large, being generally under half an inch in length and a couple of lines in breadth. It is unpolished in appearance; and though the animal within bears a resemblance to the Helices, or snails, the shell without tapers much more considerably, and in crawling the creature carries its shelter balanced on its back, directed a little to the right, at an angle of 50°, or drags it along the ground, and holds it when at rest at an angle of 45°. It is found in woods and under mossy trees, on walls and under stones. It crawls with speed, as if to justify the imputation of being very devouring, and of getting rapidly through its fortune. It is by no means rare in England. Macgillivray and Duncan

first found it in Scotland amidst the ruins of
Dunottar Castle; and it must have an affinity for
old castles, as Mr. E. J. Lowe mentions as a Mid-
land habitat Nottingham Castle yard.　　Three
examples next turn up of the chrysalis snail shell,
or pupa.　Their fancied resemblance to chrysalids
give the pupæ their name.　They are similar in their
habits to the *Bulimi*, feeding on vegetable substances; residing under mosses, herbage, or stones;
inert in continued drought, and searching mostly for
food at night, when the grass is moist.　Unlike the
Bulimus, however, the *Pupa umbilicata*, for instance,
is a slow mover, and carries its shell at the moderate inclination of 15°; and the shell, instead of
tapering like the *Bulimus*, comes abruptly to an

51.　　　　52.　　　53.　　　54.

51. Bulimus obscurus (the Dusky Twist Shell), *Müller*,
52. Pupa umbilicata (the Umbilicated Chrysalis Shell),
Draparnaud. 53. Pupa pygmœa (the Pigmy Chrysalis
Shell). 54. P. substriata (the Six-toothed Chrysalis
Shell), *Jeffreys*.

apex.　It is very minute, usually only the ninth of
an inch long and the twentieth of an inch broad.

Wherever limestone or chalk abound, there it is found, whether in the cracks of old walls, amidst ivy, under stones and the bark of trees, or on the sea-cliffs and valleys. If, however, the *P. umbilicata* be minute, the *P. pygmœa* is exceedingly minute, being not more than a line in length. By most authors this tiny shell, found, though not abundantly, in all parts of Great Britain and Ireland, both in wet and dry situations, but principally in dry, is classed as *Vertigo pygmœa.* The *P. substriata*, so called from its cylindrical, shining, polished little surface being streaked longitudinally, is much the same size, a line long, half a line broad, and though rare where it is found, is at the same time widely diffused from Cornwall, Devonshire, and Suffolk, to Preston, Lancaster, and Newcastle-upon-Tyne. And speaking of Newcastle-upon-Tyne, it ought to be mentioned that Mr. Joshua Alder, the great molluscous authority and *genius loci*, has pointed out a remarkable structure in the interior of the *pupa*, the use of which has not yet been ascertained. It consists of a raised thread-like laminar process, winding spirally round the columella, and similar lamina running spirally on the upper side of the volutes, with small flat transverse plaits at intervals in the interior.

The widely-spread *Balea fragilis* has been confounded with the *pupæ*, but it is far too tapering. It is a small, thin, delicate, shining, and rather transparent shell, of a yellowish horn-colour, and in length about the third of an inch. It is found in trunks of trees, and amidst mosses and lichens. The dark close shell, *Clausilia nigricans*, with

55.   56.

55. Balea fragilis (the Fragile Moss Shell), *Draparnaud.*
56. Clausilia nigricans (the Dark Close Shell), *various.*

which we have grouped it, is better known as the common *Clausilia ;* but it is quite a conchologist's shell, having long escaped vulgar popularity, though very generally distributed throughout Great Britain. This exclusiveness is due to its habits and colour, which render it far from easy of detection. Its length is half an inch, breadth from a twelfth to an eighth, and it inhabits old walls. The animal, as may be inferred from the shell, is very thin and slender, so much so that in motion it is incapable of raising its shell, but drags it along in the same line as its foot and neck, although when going to

repose it inclines it at an angle of 70 . It derives
its generic name from a shelly bone attached to

57   58-59.

57. Zua lubrica (the Common Varnished Shell), *Müller.*
58-59. Azeca tridens (the Glossy Trident Shell), *Pulteney.*

the columellar teeth, and termed the *clausium*, from
closing up the aperture when the animal has
retired within its habitation. The next little group,
Zua lubrica (common varnished shell), and *Azeca
tridens* (glossy trident shell), are parallel in character,
only the *Zua* is toothless, the *Azeca* ovate and
toothed in the mouth. Both inhabit close shady
wood, moss, and under stones and decayed leaves.
The one is not quite, the other about, a quarter of
an inch in length. The next group comprises the
needle agate shell (*Achatina acicula*), the shell of
the common amber snail (*Succinea putris*), and two
Physæ or bubble shells (*P. fontinalis* and *P. hypnorum*). The first is indeed minute, interesting, and
extremely delicate, having six convolutions, though
only a fifth of an inch in length; but, indeed, dead
specimens found in old Saxon coffins are more
frequent than living ones occurring amongst roots
of grass and moss. *Succinea* is from *succinum,*

amber, and *putris* means filthy; but there is nothing repulsive about the shell, which, with its variety, *S. gracilis* (slender, is found always near water,

60.　　61-62.　　63.　　64.

60. Achatina acicula (the Needle Agate Shell), *Müller.*
61-62. Succinea putris (the Common Amber Snail), *Linnæus.* 63. Physa fontinalis (the Stream Bubble Shell), *ibid.* 64. P. hypnorum (the Slender Bubble Shell), *ibid.*

either crawling on mud or damp, or attached to succulent plants. They are never found, however, in the water. Not so *Physa fontinalis*, as its name implies, *physa* (φυσαῶ), inflated or blown out, and *fontinalis*, residing in springs or fountains. Yet the creature is herbivorous, feeding on the leaves, especially of *Potamogeton*, in lakes and rivers. Beneath the water it glides along with moderate, uniform motion, produced by the undulations of its foot. In the air it advances by jerks, without protruding its tentacula: and Montagu asserts that it will sometimes let itself down gradually by a thread affixed to the surface of the water, as the *Limax* drops itself from the branch of a tree. *P. hypnorum* is found in ditches and stagnant pools in many parts

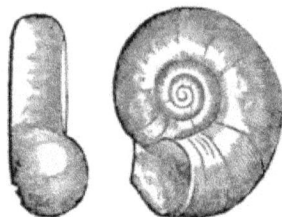

65-66.

67.

68.

69.

70.

71-72.

73-74.

75-76.

77-78.

65-66. Planorbis corneus (the Horny Coil Shell), *Linnæus.* 67. P. albus (the White Coil Shell), *Müller.* 68. P. nautileus (the Nautilus Coil Shell), *Linnæus.* 69. P. marginatus (the Margined Coil Shell), *Draparnaud.* 70. P. carinatus (the Carinated Coil Shell), *Müller.* 71-72. P. vortex (the Whorl Coil Shell), *Linnæus.* 73-74. P. spirorbis (the Rolled Coil Shell), *Müller.* 75-76. P. contortus (the Twisted Coil Shell), *Linnæus.* 77-78. P nitidus (the Fountain Coil Shell), *ibid.*

of Great Britain and Ireland. The mud is prolific
of shells; witness the Planorbis family, of which

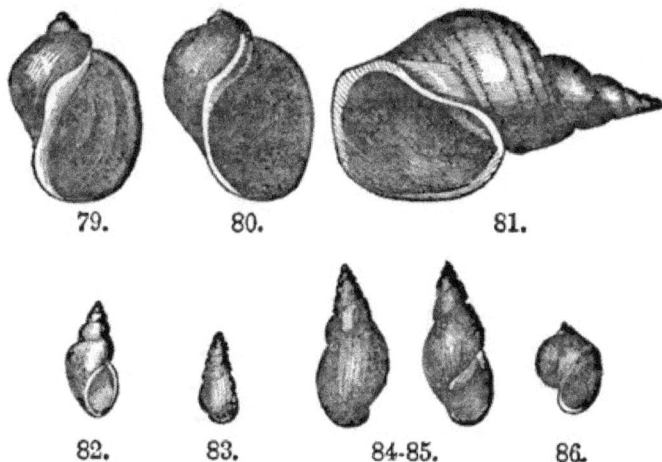

79.   80.   81.

82.   83.   84-85.   86.

79. Limnæus auricularius (the Wide-eared Mud Shell),
    Linnæus.   80. L. pereger (the Travelled Mud Shell),
    Müller.   81. L. stagnalis (the Lake Mud Shell), *Linnæus.*
    82. L. fossarius (the Ditch Mud Shell), *Turton.*   83. L.
    glaber (the Eight-Whorled Mud Shell), *ibid.*   84-85. L.
    palustris (the Marsh Mud Shell), *Linnæus.*   86. L. gluti-
    nosus (the Glutinous Mud Shell), *Müller.*

nine species are here figured. They have been
likened to the fossil Ammonites, for which they might
be taken as miniature copies. The name is hence
a kind of contradiction in terms, compounded of
the words which signify " flat " and " ball." The
largest is the *P. corneus* (Figs. 65 and 66), an inch
in diameter; the others are *P. albus*, from one-

fourth to one-fifth; *P. nautileus*, one-eighth to one-tenth; *P. marginatus*, five-eighths; *P. carinatus*, *P. spirorbis*, one-fourth; and *P. contortus*, one-fifth of an inch in diameter; respectively; whilst *P. vortex* is usually only from three to four lines, and *P. nitidus* two and a-half lines. These shells are found in the slow rivers, pools, and stagnant waters of England. The *P. vortex* does not actually reside in the mud, but on its surface; but more especially occupies the stems and leaves of plants, both in and out of the water, retiring into recesses and cavities in the banks formed by the plants or their overlying stems or leaves. In point of fact, the mud shells, *par excellence*, are the group of *Limnæcea*, of which there are given seven examples (Figs. 79—86). The family is wide-spread, the shells are fragile; and Mr. Lowe has noticed that the property of walking upside down on the ceiling, appertaining to the house-fly (*Musca domestica*), has been introduced by the Limnæus into the watery regions, " for it as easily crawls upside down on the surface of the water basking in the sun, as it moves in the ordinary manner on the surface of the mud." The lake and river limpets, *Ancylus oblongus* and *A. fluviatilis* (Figs. 87 and 88) are small breathing animals; and the minute sedge shell, *Carychium minimum* (Fig. 89), though common, is almost microscopical.

The embryo naturalist, perchance, may imagine that a complete cabinet of common shells could be easily set up; but it is not every one whose enthu-

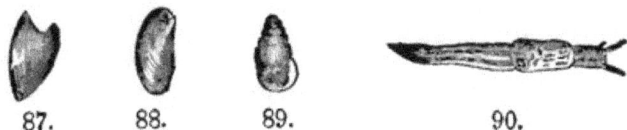

87.   88.   89.   90.

87. Ancylus fluviatilis (the Common River Limpet), *Müller*.
88. A. oblongus (the Oblong Lake Limpet), *Kightfoot*,
89. Carychium minimum (the Minute Sedge Shell),
Müller.   90. Limax brunneus (the Brown Slug), var.,
Draparnaud.

siasm would lead them to undergo the task. The man of science is well aware that he must trace out the wonders of the living creation in their native haunts, in order to their perfect comprehension; we are therefore glad to know that the York Natural History Society collects and sends out to subscribers the shells and fossils of different British districts and strata, at something like ten shillings a set.

THE END.

Crown 8vo, elegantly bound in cloth gilt. Illustrated with coloured plates and wood engravings. Price 6s.

THE AMATEUR'S
KITCHEN GARDEN
FRAME GROUND AND FORCING PIT,

A HANDY GUIDE

To the Formation and Management of the Kitchen Garden, and the Cultivation of useful Vegetables and Fruits.

BY

SHIRLEY HIBBERD.

CONTENTS.

OPINIONS OF THE PRESS.

"We recommend the book, as one from which amateurs and even professional gardeners may derive reliable information, which is the more acceptable as it is conveyed in an entertaining manner."—*Journal of Horticulture.*

"Correctly described as no mere compilation, but the result of a quarter of a century's work in gardens largely devoted to fruit and vegetable culture."—*Saturday Review.*

"A beautiful and well illustrated book, should be a very welcome addition to the library of any amateur gardener; and we very confidently recommend it to all gardeners who wish to do the right thing at the right time, and in the best and most profitable and productive manner."—*Yorkshire Post.*

"It is a complete book, dealing well and wisely with every point incident to the comprehensive subject. The author has established wide and large renown, and many are the amateurs who owe him a debt of gratitude. The publishers have done him ample justice."—*Art Journal.*

"Mr. Hibberd embodies in his work the results of his own practical experience, and the directions he gives are so simple and comprehensive that anybody who sets about it can find no difficulty in understanding and following them. A better guide than this handy volume need not be desired by the amateur gardener."—*Scotsman.*

"Mr. Hibberd is now the best known among acknowledged authorities on the subject of gardening. His present work aims wholly at utility. It shows the manner of laying out a kitchen garden to the best advantage."—*Sunday Times.*

"It shows the amateur what a kitchen garden ought to and might be, how to form one, and what to do with it when you have got it."—*Live Stock Journal.*

"An invaluable addition to garden lore, and among the best, where indeed all are excellent, of Mr. Hibberd's productions."—*Ladies' Treasury.*

GROOMBRIDGE & SONS, Paternoster Row, London.

www.ingramcontent.com/pod-product-compliance
Lightning Source LLC
Chambersburg PA
CBHW021403210326
41599CB00011B/993